MW00715233

THE BOOK OF GRASSES

THE BOOK OF GRASSES

AN ILLUSTRATED GUIDE TO THE COMMON GRASSES, AND THE MOST COMMON OF THE RUSHES AND SEDGES

BY

MARY EVANS FRANCIS

ILLUSTRATED BY H. H. KNIGHT,
ARTHUR G. ELDREDGE AND
SARAH FRANCIS DORRANCE

GARDEN CITY   NEW YORK
DOUBLEDAY, PAGE & COMPANY
1912

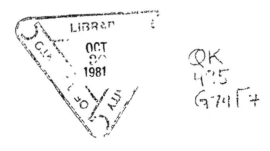

TO
S. E. F. D

" 'a babbled of green fields."

PREFACE

THE little that has been written about our common grasses has dealt chiefly with their economic value, and has been published for the agriculturist, to whom that value is paramount Or, it has, on the other hand, been too technical to be of service to the casual student of our wild flowers who has had comparatively little aid, aside from scientific and agricultural works, in recognizing the different species of this vast family.

In preparing the following pages I have intended that the descriptions, though accurate, should not obscure the beauty of the grasses with a mass of technical terms, but should be so simple that the wayfaring man who enjoys the verdure of our waysides might become more intimately acquainted with the most common plants

An important aid in recognizing the grasses will be found in the illustrations, which, made from the living plants, present not only the most noticeable characteristics of growth, but also delineations of the parts of the flowers

The technical descriptions, which follow the general descriptions, are the results of careful observations and measurements of many specimens.

The descriptions include the common grasses and the most common of the rushes and sedges found from Canada southward to Virginia, and from the Atlantic coast westward to the Mississippi River   The greater number of species given are found throughout the United States

I wish to acknowledge indebtedness to the following valuable works: "The True Grasses," by Eduard Hackle; "American Grasses," by Dr Lamson Scribner, "Grasses of North America for Farmers and Students," by William J Beal; and "Descriptive Catalogue of Grasses of the United States," by Dr. Geo Vasey.

MARY EVANS FRANCIS.

LIST OF CONTENTS

LIST OF ILLUSTRATIONS

Coloured Plates

xi

The Book of Grasses

OF GRASSES

OF GRASSES

FROM spring until late autumn grasses bloom by every wayside, and in field and meadow form the green carpet of the earth.
Widely distributed throughout all countries, and abundant even
in far-away prehistoric days, they still remain the most important
family of the vegetable kingdom, and — of all common plants the
most common — the least commonly known. Yet from the
moment when the first violet lifts its blossom to the sunlight until
in autumn the witchhazel's delicate flowers are seen above fast-
falling leaves, there is never a day when grasses are not in bloom,
and never a week in summer when a score of different species may
not be gathered

In richness and variety of colouring, above their undertone of
green, the blossoms and wind-blown anthers of the grasses rival in
beauty the flowers that the wayfaring man collects   The grace
of swaying stem and drooping leaf, the delicacy of tiny flowers
tinged in rose and purple, and the infinite variety shown in form
and colouring are lost upon those who are intent on seeking flowers
that the forests make rare.   Grasses there are, stout and higher
than one's head, and grasses so slender that their dying stems
among wayside weeds are like threads of gold, grasses whose
panicles of bloom are more than half a yard in length, and of a
colour which only a midsummer sun can burn into August fields,
grasses so stiff that winter's snow leaves them unbroken, and
grasses so tiny that their highest flower is raised but a few inches
from the soil

Nearly one thousand species are found in the United States,
nor is the study of these plants so difficult as it is thought to be
When accuracy in determining the individual species is desired, a
small microscope and a few needles for dissecting the blossoms are
all that is necessary   Even without these aids an intimate acquaintance with the grasses may be gained by observing only their
most obvious characteristics of growth, and the various forms of
flowering heads   Notice closely the grasses in a low meadow of
early summer. the dense growth of green, hastily characterized as

3

"grass," may contain many different species of this vast family, species which at a second glance are seen to have each their own distinguishing features.

Like charity the study of grasses may begin at home, and, like charity also, this most fascinating of nature studies may be carried far afield, for the grasses, most numerous of all flowering plants, we have always with us. Tree-like in the tropics the Bamboos, largest of the grass family, lift their blossoms one hundred feet and more toward the sky; in cold countries moss-like grasses cringe and cling to the frozen ground, and through the temperate regions of the globe grasses grow in luxuriance of form and colouring and supply a background of green against which the world of trees and rivers, of brooks and ledges, is placed on colours ever changing, and ever perfect Nature is continually busy reclaiming the unsightly places abandoned by man, covering with a garment of green the hillsides torn by rain, and carpeting with her "matted miracles of grass" the humble waysides.

The traditional spirit of the seasons is symbolized by outdoor colouring cool, pale tints of early spring, rose-colour of June, warm tones of August fields, and a glory of purple and gold when the summer is past and the harvest ending. In all this continuity of change, which keeps the face of Nature so new in its world-old familiarity, the grasses bear their part, and as the violet and wild geranium of spring give place to midsummer hardhack, which in turn is pushed aside by goldenrod and asters, so the passing months bring fresh grasses into bloom and mark the calendar of the year by the flowering of these common plants

It still is true, however, that

> "The world misprizes the too-freely offered
> And rates the earth and sky but carelessly."

The dandelion is less honoured than the arbutus, yet even the dandelion receives greater honour than do the early grasses, which aid in changing earth's wintry shroud to living green. Grasses yield us the earliest intimations of spring, as a faint flush of green, in harmony with the soft colours of April woods, tinges the brown hillsides before snows have ceased The first grasses are more delicately coloured than are those of midsummer when the sun burns red and purple into the tiny flowers. The green spikelets of many spring grasses depend for colour upon their lightly poised anthers of lavender and gold   Sweet Vernal-grass, Orchard Grass, and June Grass, so characteristic of spring, are succeeded by spreading panicles of Hair-grass, bayonet-like spikes of Timothy, and the richly coloured Red-top whose blossoms burn with midsummer's warmth   September has still new grasses to offer, and in this month the Beard-grasses are conspicuous, as their stiff stems at last attain a growth that will enable them to withstand snow and frost.   In many localities from fifty to one hundred different grasses may be gathered, and, although, unlike the lilies, they do not flaunt their colours garishly, yet in rose and lavender, in purple and an infinite scale of green they rest and charm the eye with their beauty from April to October, when frosts bring to them new hues of brown and yellow in which they clothe the earth until green blades again push through spring turf.

Our waysides are the accepted gardens of many plants which, having followed the path of mankind through the New World, take the highways of civilization for their own, and find abundant means for transportation as seed is fastened on passers-by, or carried by the wind along smooth pathways   Few are the grasses that cannot be found in these wayside gardens as the roads wind through fertile country, from uplands to rich meadows, or pass sandy shores, where in a variety of soils the different grasses bloom and add a mass of verdure to the border of the way   Throughout the season these common gardens of the wayside hold a constantly changing procession of grasses, a procession which begins with Low Spear-grass and Sweet Vernal-grass in April, and ends in October with the Dropseed-grasses and the Beard-grasses, although even in winter the species that remain standing may still be recognized

Rarer flowers must be sought in deep woods and in hidden places in the swamps, but the cosmopolitan grasses are fitted to take up the struggle for existence wherever the seed chances to fall   Dean Herbert rightly says that "plants do not grow where they like best but where other plants will let them."   By waysides we may see this struggle in its intensity as a dozen species strive for the same plot of ground and grow in tangles that include low cinquefoil and tall briars   The strife is always most intense between individuals of the same species, and here the grasses grow in profusion, occupying each inch of space, pushing

5

out into deserted country roads, and spreading far and wide by means as interesting as ever the more noted flowering plants employ

Bur-grass, with its thorny seed-burs, catches on passing objects and thus secures free portage to new fields, Terrell-grass by thick, corky scales floats its seed upon the streams near which it grows, Beach Grass defies the sand to bury it and is found at the tops of the highest sand-dunes, with whose rise it has kept pace, the long roots of the grass penetrating to the base of the dune, and Couch-grass, sending sharp-pointed rootstocks rapidly through the soil, is a veritable "land-grabber"

Where the purslane and poppy produce a multitude of seeds from every flower, each blossom of the grass ripens but one, yet so richly stored is this with nutriment, and frequently so well protected against germination under unfavourable conditions, that the one seed may be worth many of those less perfectly equipped, since, in the process of evolution, diminution in the number of seeds is accompanied by an increase in the effectiveness of those that remain

The twisted awns of certain grasses — e g , Sweet Vernal-grass and Wild Oat — show one of the most interesting mechanisms seen in the vegetable world These awns, or bristle-like appendages of the grass flower, are extremely sensitive to atmospheric changes, and by their peculiar structure aid in burying the seed beneath the surface of the soil. In Sweet Vernal-grass the scale, to which the ripened seed adheres, bears a brown awn, bent and twisted near its middle, and beset with minute, upward-pointing hairs on its basal part. Such awns are strongly hygroscopic and during cold or dry weather remain tightly twisted, thus holding the seed where it chances to be Under the influence of moisture the awn untwists and by its rotation drives the fallen seed slowly but surely beneath the soil Although dry weather may follow, causing the awn to become twisted again, the upward-pointing hairs catch on particles of earth or grass and, holding the seed down, prevent it from being drawn up. Thus it lies ready for the next shower when the awn pushes the seed farther into the earth This peculiarity of structure is easily observed without the aid of the microscope. If a few of the ripened seeds be laid upon the moistened palm of the hand they will immediately begin to move, as if alive, and the rotating of the awn may be plainly seen.

6

Interesting experiments have been made whereby it has been seen that in sand, alternately wet and dried, the awns of certain grasses will bury the seed several inches beneath the surface

Each locality shows characteristic grasses, and as in a short walk we pass from low meadows to dry hillsides we find new species to excite fresh interest. On sea beaches we look for the long, gray-green leaves of Marram Grass, or Beach Grass, for spreading clumps of Sea-beach Panic-grass, for the dark, wiry stems of Fox-grass, and for rigid-leaved grasses of hot sands Salt marshes show dense jungles of reed-like grasses, Creek Sedge, Salt Reed-grass, and the tall Reed. Dry hillsides are covered in spring by Wild Oat-grass and Wavy Hair-grass, where later Purple Finger-grass, Sheathed Rush-grass, and stiff Beard-grasses will bloom In dry fields we look for the low growth of the smaller Panic-grasses, for the slender, one-sided spikes of Field Paspalum, and for wide-spreading panicles of Purple Eragrostis Borders of woodlands offer Poverty Grass, Black Oat-grass, and Muhlen-bergias, while in deep woods we search for shade-loving grasses, the tall, slender Bottle-brush Grass, the lower Mountain Rice, and the Nodding Fescue Marshy meadows are full of interest to the student of grasses. Reed Canary-grass with broad, blue-green leaves borders narrow brooks, and nearby the Blue-joint Grass, slender and stiff, rises bearing narrow, deeply coloured panicles, graceful Manna Grasses fill the marshes of early summer, and later the rough leaves and stems of Rice Cut-grass form tangled masses in low grounds By river-borders grows the great Gama Grass whose leaves are so broad as to resemble those of our cultivated corn, and in wet soil, also, is found the tall Indian Rice on which the reed-bird feeds. A country dooryard of an acre may show more than a dozen different grasses, while in the garden near half a score of other species invade the cultivated land as weeds A large collection of grasses, preserved either as herbarium specimens or in the more artistic impression prints made upon photographic paper, may be gathered in a short time, and differences perhaps little noticed by the casual observer will seem marked indeed to the student who at the close of a summer's study will deem it as unpardonable to mistake one of our common grasses for another as to mistake an elm for an oak

Corn, wheat, oats, the day of the first cultivation of these cereal grains long antedates history, and how seldom is it realized that they are grasses   Vergil and Columella wrote long ago of the care of meadows and fields   Indeed the word cereal stands as an article of faith in the goddess Ceres, who searched with torches for the grain carried off by winter frost, and on finding the seed raised it to its flower once more   Bertha was the Ceres of German mythology, and winds and rains affecting crops were believed to be under her control.   Corn-spirits there were which were symbolized under the forms of wolves and goat-legged creatures, similar to classic satyrs.   To the older peasantry of Germany and Russia these corn-spirits still haunt and protect the fields which show the "Grass-wolf" or "Corn-wolf" to be abroad when the wind, as it passes, bends the grass and the ripening grain.   The last sheaf of rye is occasionally left afield as shelter for the "Roggen-wolf," or "Rye-wolf," and it is not long since the Iceland farmer guarded the grass around his fields lest the mischievous elves, hiding among the grasses, and ever waiting to harm him, should invade his cultivated land

In old herbals the word *grass, gres, gyrs* meant any green plant of small size, and though we have restricted the meaning of the word it still is carelessly applied to a multitude of sedges and rushes which in manner of growth and form of flower differ markedly from the true grasses.   To the casual observer the grasses are but "grass," and to few is their diversity, their beauty, and their value apparent.   We are blind to the infinite variety shown by Nature in these common plants, of which we often know scarcely more than do the cattle that feed upon them, yet on no other family of flowering plants does the beauty of the green earth and its adaptation as a home for man so largely depend

UTILITY OF GRASSES

THE MOST IMPORTANT FAMILY OF THE VEGETABLE KINGDOM

"And he gave it for his opinion, that whoever could make two ears of corn, or two blades of grass, to grow upon a spot of ground where only one grew before, would deserve better of mankind, and do more essential service to his country, than the whole race of politicians put together "—*Gulliver's Travels*.

CAN one imagine the world grassless — a barren waste? The shifting soil, exposed to the elemental workers, wind and water, could offer no sure abiding place for man, since, lacking a tenacious network of grass roots firmly binding the soil, the road of to-day might be obliterated to-morrow, and the loftiest building gradually buried beneath wind-blown sand.

As soil-binders the grasses performed a leading part in the important task of rendering the globe habitable to the human race, and still sending their roots far and wide through the surface of the ground the grasses form a turf which holds in check the destructive forces of wind and rain, and gives secure anchorage not only to the lower growth of plants but also to trees and shrubs

Grasses were abundantly developed in prehistoric days, as numerous remains of grass-like leaves attest, and since the earliest tribes chipped rude implements for cultivating the soil, or for their use in war, the grasses have exceeded in importance to mankind any other family of the vegetable kingdom.

The green herbage of meadow and pasture is the chief food of domesticated animals, and in this country the value of hay alone exceeds that of any other crop except corn, which, be it remembered, is itself a grass Even the salt marshes yield their hay, and in New England pastures, where rocks seem as numerous as grass blades, sheep crop the wiry grasses of dry hillsides

A noted grass-garden was owned in Woburn a century ago by the Duke of Bedford, and in this garden George Sinclair carried on valuable researches of which he wrote in his "*Hortus Gramineus Woburnensis*." Tirelessly were the experiments made, in cultiva-

11

ting the grasses, in drying them, in dissolving their soluble parts, in evaporating the solution, and finally in submitting the residuum to chemical analysis

Grass stems contain a large amount of silica, and in such seed as that of the species known as "Job's Tears" the hardness due to a silica deposit nearly equals that of agate   Minute particles of silica in the outer cell walls serve in keeping grass stems firm and erect, and if we carefully burn the vegetable matter from one of these stems a perfect skeleton of the structure is left.   It is said that wheat straw, without the addition of other material, may be melted into colourless glass, and that barley melts into glass of topaz yellow.

The varied form and texture of the grasses adapt them to many uses, and even the common grasses of our northeastern states have been made into ropes, mats, paper, baskets, and many fine-plaited articles   Fragrant fans of dark-coloured fibres are made in India from the aromatic rootstocks of a grass, and the entire plant is woven into screens which, when dampened and placed in a current of air, perfume the breeze.   Lemon Grass and Ginger Grass, natives of tropical Asia, yield oils strongly scented, as their names imply, and the rootstock of a grass in South America is sometimes used as a substitute for soap

A few grasses have been used medicinally, and have been cultivated for medicinal purposes   But it is as food for man, and for the domesticated animals on which he is most dependent, that the grasses have attained their highest importance, and it is on them largely that the great human family is fed to-day.

The world has seemed to draw a line between the grasses of the fields and those plants that produce well-filled heads of cereals, and has ceased to regard the latter as grasses   Yet the useful grains — corn, wheat, rye, barley, rice, and oats — belong to one family, and are but grasses that have been brought by man to a superior degree of excellence   Rice and wheat have been cultivated from time immemorial, and although a century ago wheat was wheat, yet to-day new strains have been developed which grow where in older days the grain could not have been raised

Indian corn originated in tropical America, and is one of the few cereals whose native condition is known.   It had attained a wide distribution when this country was discovered, and the grain must have been in use in very ancient times.   Early explorers

found the Indians cultivating corn with primitive implements —
hoes made of a sharpened stone or the shoulder blade of a moose
— and even then the seeds were described as "somewhat bigger
than small peason," while later the Pilgrims could boast the cultivation of varieties of which "the graine be big." Botanically, corn is
one of the most interesting of the grasses and is very unlike those
found in daily walks through the country. The stamens of corn
are in ornamental spikes which terminate the stems, while below,
on spikes which are borne in the axils of the leaves, are the fertile
flowers These are densely crowded on a thickened rachis,
commonly known as the corn-cob, and are covered with husks
which are the sheaths of abortive leaves and which have the leaf-
blades more or less developed The flowering scales and palets
are found in the chaff covering the cob, and the silken "tassels"
at the summit of each ear are elongated pistils Aerial roots,
thrown from the lower nodes, serve as prop-roots, supporting the
stem, and imitating in a small way the growth of a few other
tropical plants

Sugar Cane is also a grass that has been brought from the
wilderness and has been made to pay the toll of usefulness which
man would fain exact of all vegetation

In warmer countries the great Bamboos, which are but grasses
of a larger growth, are utilized as shelter, clothing, and food Of
these giant grasses houses are built which may be entirely furnished
with articles made of Bamboo, and the household, wearing jackets
and hats made of the same material, may gather tender shoots of
the plant for use as a vegetable A small section of the stem forms
a cup, and a larger section a pail; paper and ropes are manufactured from the plant, umbrellas and exquisite boxes are made of
the split internodes, and intricate appliances for spinning are
fashioned entirely of Bamboo. And these are but a few of the
uses that the several species of these grasses serve. Indeed, a
complete list of articles made of Bamboo would be a catalogue far
too long for insertion in these pages.

Impenetrable "canebreaks" of the South are formed of two
grasses similar to the Bamboos, though smaller in growth The
stout, jointed stems of the more southern species (the Large Cane),
are used for fishing rods and are made into canes and pipes. As
thatching the stems form a strong and serviceable shelter, and
when split are woven into baskets and mats. The Small Cane

13

(*Arundinària técta*) grows as far north as Maryland, and by streams and river banks forms evergreen thickets from three to twelve feet in height   A reed of southern Europe and Palestine belongs to a closely related genus, and from this grass the heroes of Homer are said to have made their arrows and with it to have thatched the tent of Achilles   Pan-pipes, such as Orpheus might have used in charming the Dryads from their leafy shelters, were also made from the smooth stems of this reed.

Of all flowering plants the grasses are the most widely distributed, and innumerable are the ways in which they have served mankind since, in the story of Eden, the earth brought forth these common plants as the first of its flowers.

STEM, LEAF, AND FLOWER

STEM, LEAF, AND FLOWER

Roots — Many grasses spread in all directions by strong runners, or rootstocks, as they are called, which are, in reality, underground stems These runners differ from the true, fibrous roots in consisting of a succession of joints from which upright stems arise, and from which true roots penetrate the soil and anchor the rootstock as it stretches far from the parent plant. Such grasses rapidly take possession of the ground, and as the rootstocks, interlacing in endless network, are thickened with a large amount of nourishing material, these grasses are enabled to endure drouth and unfavourable seasons Rootstocks of the more vigorous grasses grow many feet in a season, and the thorny, needle-like points of the growing ends often penetrate tubers and roots.

Grasses that develop only fibrous roots grow more frequently in tufts and bunches Of these grasses the Bitter Panic-grass and the common Orchard Grass are examples Perennial grasses are more numerous than annual grasses and may usually be recognized by the presence of sterile shoots growing from the lowest joint of the stem The greater number of perennial grasses bloom earlier than do the annual grasses, though some perennials are late in flowering, as, for example, the Beard-grasses and the Muhlenbergias

Stems — Grass stems are divided by joints into internodes (the space between the base of one sheath and that of the next), the point from which each sheath rises being called a node Although nearly all grasses, with the exception of Indian Corn, Sugar Cane, Gama Grass, and the Beard-grasses, possess hollow stems, which are always closed at the nodes, the rootstocks are usually solid, and the internodes of the young stems are also solid, becoming hollow by the separation of their original pith cells, which cease to grow The nodes remain solid and, being darker in colour, appear as bands encircling the stems

Nodes perform an interesting and important function in raising stems that have been bent down. Internodes play little or

17

The Book of Grasses

no part in such service, but if one notices grasses that have been beaten to earth by heavy showers, it will be seen that the lower nodes have lengthened on the side turned earthward, and that the stems are thereby bent upward at sharp angles.

Sheaths.— The broad, basal portion of each grass leaf is known as the sheath, and, encircling the stem, is an important protection to the growing internode. Each sheath is usually split, or open, on the side opposite the leaf, and the edges of the sheath overlap or partly encircle the stem a second time. In successive internodes these edges lap alternately to right and left, and the rolling of young leaves also alternates in like manner. In a few grasses — e. g., Kentucky Blue-grass and Orchard Grass — the sheaths are perfectly closed at first and are split only as the inflorescence forces its way up.

Ligule.— At the summit of the sheath is usually a thin membrane, the ligule, which closely embraces the stem and appears as an additional upward growth of the sheath or a continuation of its delicate lining. In each species the ligule is constant in form, sometimes consisting of but a tiny ring or frequently appearing as a fringe of hairs.

Leaves.— Grass leaves are borne alternately on opposite sides of the stem, in what is technically called the two-ranked growth.

The leaves of a few tropical grasses approach in form those of other families of plants, and in certain species a true petiole is inserted between sheath and blade, but in temperate regions grass leaves vary only in width and length and are always "grass-like," showing long, parallel veins

18

The leaves of many grasses are twisted (to the right in some species, to the left in others), as twining plants twist with or against the sun. It is said that the leaf blade in a few grasses is sensitive, and slowly folds together when briskly rubbed.

In dry weather and in dry soil it will be noticed that the leaves of certain grasses are rolled tightly, becoming involute, as it is called. As the cells on the upper surface of the leaf lose their moisture and contract under a burning sun the edges of the leaf curl inward until the stronger cuticle of the lower surface is outermost, and thus an added protection is given against an excessive loss of moisture. The response to the external stimuli of heat and cold, of light and darkness, in the vegetable world is exquisitely delicate. In the growth of plants, in their "sleep" at night, and in their many so-called "adaptations" to varying conditions, the student may read the life of Nature in an ever open book.

INFLORESCENCE

A *Spike* is formed when the spikelets are apparently sessile on the main axis — e. g., Couch-grass.

A *Panicle* is formed when the spikelets are on secondary or further-divided branches — e. g., Orchard Grass and Old Witch-grass.

The *Rachis* is that part of the stem on which the spikelets or spikelet-bearing branches are borne.

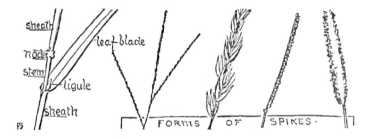

sheath
node
stem
ligule
sheath
leaf blade

FORMS OF SPIKES.

The small flowers of the grasses bear little resemblance, at first glance, to the distantly related lilies; yet if some of the lilies that bloom in spikes were to crowd their flowers more and more, and were to reduce their petals to mere scales, such plants would be well on the way toward a grass-like appearance. The three stamens of many grasses suggest the characteristic, three-parted form of the true lilies, while the flowering scale and palet of each grass blossom are a reminder of the lily calyx, the two green keels of the palet suggesting that two divisions of the calyx have been merged in one.

Our wind-fertilized flowers are represented chiefly by the grasses and sedges, and by early blooming trees and shrubs. Such flowers are small and produce no nectar. They have little fragrance, and their chief colouring frequently appears in the large anthers which are so hung on hair-like filaments as to shake out pollen grains on every breeze.

Spikelets.— The flowers of grasses are borne in spikelets which vary in size and which are composed of one, several, or many flowers. The short stem, on which the flowers of a spikelet are placed, is known as the rachilla; this is sometimes prolonged, and, under the microscope, may be seen as a tiny thread lying outside the uppermost flower. Spikelets are arranged in spikes or panicles. In bloom the lower flowers of the spikelets bloom before the others, as the spikelets bloom from below upward, but in panicles the uppermost spikelets are the first to open, since the flowering-heads bloom downward, and often the upper branches of a panicle are widely spread with open flowers while the lower branches remain erect and closely appressed to the stem.

Narrow Panicle

Spike-like Panicles

Scales.— Instead of flowering-leaves of sepals and petals the grasses show bracts, called scales, or glumes, surrounding each flower. The two lower scales of each spikelet are usually empty, and in the axil of each succeeding scale (except sometimes the

20

Rachis — Pedicels
Branches — Spikelets
Flowering Scales
Outer (empty) scale — Rachilla
several-flowered spikelet

Palet — Flowering scale
Outer (empty) scales
1 flowered spikelet

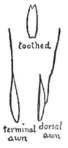

Flowering scale — Palet
Pistil
Stamens
stigmas
Lodicules — Filament
anther

Scale forms

obtuse acute

keeled 3-nerved

toothed

terminal dorsal
awn awn

uppermost) a flower is borne. Scales which enclose a flower are termed flowering scales. These exhibit many interesting peculiarities in their structure, often bearing a bristle-like appendage, called an awn, which is considered by botanists to be a modified leaf-blade. Such awns are straight, bent, or twisted, and either terminate the scales, when they are known as terminal awns, or are borne on the backs of the scales, when the awns are said to be dorsal; that part of the scale below the awn representing the sheath of a leaf, while the portion of the scale above the awn corresponds to the ligule. A flowering scale is said to be keeled when it is flattened and folded so that its two edges are brought near together and the mid-vein is prominent as a ridge on the back of the scale. When the veins of a scale are conspicuous the scale is said to be three-nerved, five-nerved, seven-nerved, or nine-nerved, according to the number of prominent veins.

Palet.— Opposite the flowering scale, and with it enclosing the flower, is an awnless scale, called the palet, usually thin in texture, and two-nerved, showing two green keels. The palet may be minute or lacking, as in certain of the Bent-grasses, or it may exceed the flowering scale in length, as in Sharp-scaled Manna-grass.

Lodicules.— At the base of the flower, within its scales, are usually two (rarely three) minute, thin, and translucent scales, termed lodicules. These will rarely be noticed save at the time of flowering, when, for a short time, they are swollen with sap, and, by pressing the flowering scale and palet apart, cause the opening of the blossom. Lodicules soon wither, and in some grasses are

21

lacking; in such the spikelets remain closed and the stamens and pistils protrude from the summit of each blossom

Stamens and Pistils. — The majority of our grasses bear perfect flowers, consisting of stamens and pistils, although some species are monœcious, as are Gama Grass and Indian Corn, which bear stamens and pistils in separate flowers on the same plant, and a few grasses are diœcious, as is Salt-grass, whose stamen-bearing and pistil-bearing flowers are on separate plants.

There are one to six (usually three) stamens whose very slender filaments bear two-celled anthers These are lightly attached near their middle to the apex of the filament, and, trembling in the wind, easily discharge the smooth, round pollen cells The stamens elongate rapidly and exhibit the most rapid rate of growth known in flowering plants. Although many of the pollen cells must fail of their mission and be carried by the wind to fall fruit-lessly upon leaves and stones, Nature provides a vast quantity of pollen to ensure the fertilization of sufficient seed. It has been estimated that a single anther of Rye contains no less than twenty thousand pollen cells The greater number of spring grasses have larger anthers than those of midsummer, but brilliant colours, ranging from pale yellow to orange and crimson, and from lavender to deep purple, appear in the anthers at all seasons The one-celled, one-seeded ovary bears one to three (usually two) styles whose feathery stigmas often show conspicuous colour

Seeds — Grass seeds are richly stored with nutriment and have great vitality; they are also well adapted to wide distribution Scales adhering to the seeds buoy them so that they are easily carried by the wind or along the surface of running water The seeds of a few grasses are sticky when wet and adhere to passing objects. Ripened panicles of Purple Eragrostis, of Old Witch-grass, and of certain other grasses are driven as tumble-weeds across the fields and scatter their seeds along the way. The awns of many grasses are rough, catching on passers-by and travelling long distances In high mountains, where the ripening of seed is uncertain, entire spikelets are sometimes transformed into leafy shoots, provided at the base with rudimentary roots, which, as the spikelets fall, take root and grow The methods which the grasses have developed to ensure to new generations transportation to new fields are many, and to them may fitly be applied the comment of Darwin on cross-fertilization devices "They transcend in an incomparable degree the contrivances and adaptations which the most fertile imagination of the most imaginative man could suggest with unlimited time at his disposal."

THE COMMON GRASSES

A CALENDAR OF THE COMMON GRASSES ACCORDING TO THEIR SEASONS OF MOST ABUNDANT BLOSSOMING

APRIL 15th to June 15th:

Low Spear-grass (*Poa annua*)	195
Sweet Vernal-grass (*Anthoxanthum odoratum*)	94
White-grained Mountain Rice (*Oryzopsis asperifolia*) . .	99
Slender Mountain Rice (*O. pungens*) . .	99
Downy Brome-grass (*Bromus tectorum*) .	222
Meadow Foxtail (*Alopecurus pratensis*)	112
Orchard Grass (*Dactylis glomerata*) .	188
Meadow Oat-grass (*Arrhenatherum elatius*) .	142
Black Oat-grass (*Stipa avenacea*) . .	100
Kentucky Blue-grass (*Poa pratensis*)	195
Canada Blue-grass (*P. compressa*)	197
Early Bunch-grass (*Sphenopholis obtusata*)	134
Meadow Sphenopholis (*S pallens*)	134
Slender Sphenopholis (*S nitida*)	134
Velvet Grass (*Holcus lanatus*)	133
Silvery Hair-grass (*Aira caryophyllea*)	132
Wild Oat-grass (*Danthonia spicata*)	143
Narrow Melic-grass (*Melica mutica*) .	181

June 15th to July 20th·

Reed Canary-grass (*Phalaris arundinacea*) .	90
Vanilla Grass (*Hierochloe odorata*) .	97
Long-awned Wood-grass (*Brachyelytrum erectum*)	108
Brown Bent-grass (*Agrostis canina*) .	115
Rough Hair-grass (*A hyemalis*) . . .	115
Black-grained Mountain Rice (*Oryzopsis racemosa*)	99
Sheep's Fescue (*Festuca ovina*)	217
Slender Fescue (*F octoflora*) .	217
Nerved Manna-grass (*Glyceria nervata*) .	206

ACCORDING TO LOCATIONS

A LIST OF GRASSES ARRANGED ACCORDING TO LOCATIONS

(In each division the grasses are given in their order of flowering.)

Grasses found in cultivated land.

Low Spear-grass (*Poa annua*) 195
Downy Brome-grass (*Bromus tectorum*) 222
Squirrel-tail Grass (*Hordeum jubatum*) 238
Chess (*Bromus secalinus*) 229
Smooth Brome-grass (*B. racemosus*) 230
Wire-grass (*Eleusine indica*) 165
Couch-grass (*Agropyron repens*) 232
Large Crab-grass (*Digitaria sanguinalis*) . . . 56
Small Crab-grass (*D. humifusa*) 56
Yellow Foxtail (*Setaria glauca*) 77
Green Foxtail (*S. viridis*) 77
Cockspur Grass (*Echinochloa crusgalli*) 62
Old Witch-grass (*Panicum capillare*) . . . 71
Meadow Muhlenbergia (*Muhlenbergia mexicana*) . . . 105

Grasses found in fields and meadows:

Low Spear-grass (*Poa annua*) 195
Sweet Vernal-grass (*Anthoxanthum odoratum*) 94
Meadow Foxtail (*Alopecurus pratensis*) . . 112
Orchard Grass (*Dactylis glomerata*) 188
Meadow Oat-grass (*Arrhenatherum elatius*) . . . 142
Kentucky Blue-grass (*Poa pratensis*) 195
Canada Blue-grass (*P. compressa*) 197
Velvet Grass (*Holcus lanatus*) 133
Brown Bent-grass (*Agrostis canina*) 115
Sheep's Fescue (*Festuca ovina*) 217
Meadow Fescue (*F. elatior*) 217
Timothy (*Phleum pratense*) 111

33

Grasses found on sands and salt marshes

Grasses found in woodlands.

Grasses found in dry soil:

Grasses found in moist soil:

KEY TO THE GRASSES

KEY TO THE GRASSES

In the illustrated description of the grasses the species follow one another in the order given in modern botanical works, such arrangement being based on the characteristics of spikelet and flower. Technical descriptions are given of the more common species of a genus. The general descriptions include other species and note their chief characteristics.

The key, being intended for use in the field, is based on the characteristic form of the flowering-head. Somewhat arbitrarily the terms used in describing the form of the infloresence have been restricted to *spike* and *panicle*, omitting the word raceme and, instead, using the term spike to include any flowering-head in which the spikelets have the appearance of being placed directly on the main axis of inflorescence. Thus, the apparent form of the flowering-head is noted and that of Timothy is given as a spike, since it has that appearance, although it is, in fact, a spike-like panicle, the spike-like racemes of Bur-grass, Gama Grass, and others are given as spikes

The generic and specific names are those given in Gray's "New Manual of Botany" (Seventh Edition)

In the key and the technical description in the following pages the measurements are in feet, inches, and lines

The symbol ' is used after figures to indicate inches, and the symbol " is used to indicate lines   A line is the twelfth part of an inch, hence 3" equal one quarter of an inch, 8" equal two thirds of an inch, etc

ARTIFICIAL KEY

Based on the more noticeable characteristics of the inflorescence·

Inflorescence consisting of spikes or spike-like panicles — I

Inflorescence consisting of short panicles, one to four inches in length — II.

Inflorescence consisting of longer panicles, size variable, usually more than four inches in length — III.

II — INFLORESCENCE CONSISTING OF SHORT PANICLES, ONE TO FOUR INCHES IN LENGTH

III — INFLORESCENCE CONSISTING OF LONGER PANICLES, SIZE VARIABLE, USUALLY MORE THAN FOUR INCHES IN LENGTH.

PAGE

Panicle 5′-8′ long, branches slender; spikelets few, narrow; flowering scale bearing 3 spreading awns

Sea-beach Aristida   105

Panicle 3′-10′ long, spikelets flat; scales 2   Leaves very rough Grasses of wet grounds   White-grass   .   90

Rice Cut-grass   90

Panicle 5′-15′ long, many-flowered, scales 3, stamen 1   Tall, leafy grasses of woods and swamps

Wood Reed-grass   131

Slender Wood Reed-grass   131

Panicle pyramidal, scales 4, lowest scale small, flowering scale white, porcelain-like, shining   Several species of Panic-grasses   61

Panicle 2′-10′ long, pyramidal, open, spikelets small, about 1″ long. Palet one third as long as flowering scale.

Red-top .   120

Palet minute or wanting   Brown Bent-grass .   116

Thin-grass . . .   116

Panicle 3′-10′ long, pyramidal, open, spikelets small.

Panicle many-flowered, base enclosed in upper sheath

Sand Dropseed   .   115

Panicle very delicate   Gauze-grass   115

Panicle 6′-24′ long, pyramidal; branches hair-like, long and widely spreading, spikelets small   Rough Hair-grass   .   120

Panicle 6′-18′ long, branches hair-like, widely spreading, spikelets small on hair-like pedicels, flowering scale bearing a straight, terminal awn 3″-9″ long.   Long-awned Hair-grass . .   108

Division B.   Spikelets more than 1-flowered

Panicle 2′-8′ long, narrow, yellowish, spikelets 2-flowered, scales 4, flowering scale of upper flower bearing a bent and twisted awn.

Marsh Oats . .   135

Panicle 4′-10′ long, narrow; spikelets 2-flowered, scales 4, lower flower staminate, its scale bearing a dorsal awn, upper flower perfect and awnless   Meadow Oat-grass   143

Panicle 1′-6′ long, narrow; spikelets 6-12-flowered, flowering scales short-awned.   Slender Fescue   .   221

Panicle 2′-8′ long, contracted, spikelets 2-3-flowered, 1st scale narrow, acute, 2nd scale much broader, obtuse   Grasses of early summer   Early Bunch-grass   .   135

Meadow Sphenopholis   .   136

Slender Sphenopholis   .   135

Panicle 2′-12′ long, spikelets 3-10-flowered, 2½″-6″ long.   Panicle narrow and contracted after flowering

Meadow Fescue   222

Panicle slender, nodding, branches few

Nodding Fescue .   .   218

Panicle erect, short, branches spreading

Red Fescue .   221

43

ILLUSTRATED DESCRIPTIONS
OF THE GRASSES

"Where thou with grass, and rivers, and the breeze,
And the bright face of day, thy dalliance hadst."

GAMA GRASS

THE day when this grass is first seen, and is recognized as a member of the same family whose smaller species are commonly trodden under foot, is a day unforgotten by the nature-lover With stems more than shoulder-high, with leaves so large as to resemble Indian Corn, and with thick spikes of oddly formed blossoms, the Gama Grass, as it grows in low meadows and along streams, is one of the largest and most remarkable grasses of the Eastern States.

The coarse, branching stems rise from stout rootstocks and, unlike those of the majority of grasses, are solid, being filled with pith The blossoming spikes are peculiar in form, the stamens and pistils are in separate flowers, and in midsummer long, orange-coloured anthers clothe the upper portion of each spike, while for a short time feathery stigmas of dark purple hang from the pistillate flowers below. These fertile flowers are deeply embedded in boat-shaped cavities which are closed by hard and shining scales, and as the upper portion of the spike soon falls, the thick basal part is left, and easily breaks into short joints each containing a seed. Although smooth and shining, these seed-capsules lack the symmetry of form and the agate-like surface which characterizes the fruit of "Job's Tears," a closely related species which is occasionally cultivated as an ornamental grass and whose seeds are sometimes used for rosaries

TECHNICAL DESCRIPTION

Gama Grass. *Tripsacum dactyloides* L

Plant perennial, from stout rootstocks

Stem 3-8 ft tall, solid, stout, erect, branching Leaves 1 ft long or more, 6"-18" wide

Spikes 2-4. 4'-9' long at summit of main stem, solitary spikes on the branches Spikelets of two forms, upper part of spike composed of 2-flowered, staminate spikelets about 4" long, outer scales obtuse; 1-flowered pistillate spikelets below deeply imbedded in the rachis, outer scale of pistillate spikelets hard and shining, enclosing the

49

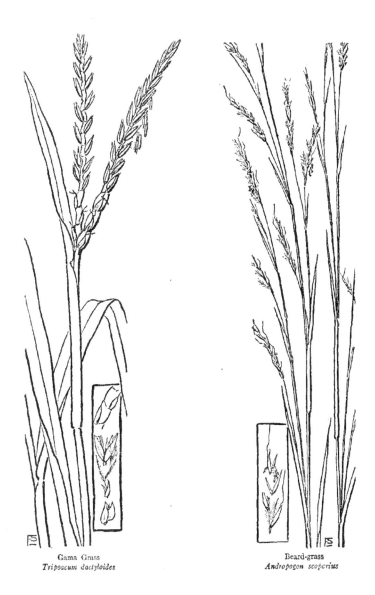

Gama Grass
Tripsacum dactyloides

Beard-grass
Andropogon scoparius

50

CHARACTERISTIC TYPE OF BROOM-SEDGE lands in central Texas ... he blossoming

recess in which the flower is embedded. Stamens 3, anthers orange colour, large. Stigmas purple, long.

Moist soil, swamps, and borders of streams. June to September.

Rhode Island to Florida, Texas, Missouri, and Kansas.

BEARD-GRASS, BROOM SEDGE, FORKED BEARD-GRASS, AND BUSHY BEARD-GRASS

When the royal purple and gold of asters and goldenrod paint the waysides, and mark the turning toward harvest of the tide of midsummer, the Beard-grasses also appear as the vanguard of autumn and show the advancing season as surely as do the more brilliant flowers. In every state, from coast to coast, these grasses grow, characteristic of dry, sandy soils, and easy of recognition. The species are looked upon with little favour in the East, but in Western pastures, on prairies and ranges, the Blue-stems, as these plants are locally called, yield a valued herbage.

Tufts of Beard-grass, the most common of the genus in Eastern States, are frequently seen by waysides, in sandy fields, and near the borders of dry woods. This grass, sometimes known as Indian Grass or Little Blue-stem, is late in starting and the leaves, often tinged with red and bronze, are seldom noticeable until June. In July the slender, rigidly erect stems appear, usually bluish purple in colour and at last fringed with small solitary spikes of hairy blossoms which hang to the winds their orange and terra-cotta anthers and purple

Forked Beard-grass
Andropogon furcatus

stigmas. Not until September, however, is the plant in its greatest beauty, as the spikelets at maturity change to tiny silvery plumes adorning the ripened and richly coloured stems.

In similar locations, though less common in the North, is the Broom Sedge (*Andropògon virgínicus*), which may be distinguished by an examination of the spikes; those of this species being borne in pairs or several together. In the South this grass is much stouter, and on mountainsides and in lowlands it covers the fields with its rank growth. Aside from its value to the farmer in early summer, Broom Sedge, as its name indicates, finds later a more humble use in the household. Great handfuls of the stout stems are tied together, and when the hairy spikelets are beaten out, and the slender tips cut off, a serviceable, brush-like broom is ready for immediate use on hearth and floor.

Stiff, brown groups of Beard-grass and Broom Sedge remain standing through all the winter months, and are as easily recognized in March as they were in the preceding summer. Brilliant colours are rare when Nature is clothed in the dull brown of faded leaves, but these grasses, beneath their neutral tones, hold a colour more striking than in summer. On a wintry day strip from the stem one of the dry sheaths. The inner surface glistens with colour varying from pale yellow to copper colour and bright orange-red, while in a closely related species of the South (Johnson Grass, *Sórghum halepénse*) the long sheaths are lined with glowing crimson.

Forked Beard-grass blooms by fences and hedges in early autumn. The tall stems, rich in colouring, are surmounted by short, spreading spikes of reddish brown or purple, and by this finger-like inflorescence the grass is easily recognized.

Bushy Beard-grass (*Andropògon glomeràtus*) is found in damp soil from New York southward. It is rarely more than three feet tall, and as the branches which bear the spikes are elongated the stems are crowned with dense, terminal panicles of hairy blossoms.

Aromatic perfumes are prepared from certain foreign grasses of this genus. Citronella oil is distilled from a species of Hindostan, and the roots of another are woven into the "Vessaries," or fan-screens, which, when dampened and hung in a current of air, before door or window, perfume and cool the house.

Beard Grass. Little Blue-stem. *Andropògon scopàrius* Michx

Perennial, usually tufted

Stem 1-4 ft tall, solid, slender, erect   Ligule less than 1" long   Leaves 4'-10' long, 1"-3" wide

Spikes numerous, 1'-2' long, loosely flowered, solitary, terminal and along the stem   Spikelets in pairs on a hairy rachis, hairs dull white, conspicuous, 1 spikelet of each pair sessile, perfect, 1-flowered, about 3" long, bearing a twisted, bent awn 5"-7" long, the other spikelet of the pair sterile, borne on a hairy pedicel and reduced to an awn-pointed scale   Stamens 1-3, anthers terra-cotta or yellow

Dry soil   July to October

New Brunswick to Alberta, south to Florida, Texas, and southern California

Forked Beard-grass. Big Blue-stem. *Andropògon furcàtus* Muhl

Perennial

Stem 3-6 ft tall, stout, erect   Ligule 1" long or less   Leaves 6'-16' long, 2"-6" wide, roughish

Spikes 2-5 purplish, 2'-5' long, rather thick and rigid, spreading from summit of culm and lateral branches   Spikelets in pairs on hairy rachis, hairs short, 1 spikelet of each pair sessile, perfect, 1-flowered, 4"-5" long, bearing a loosely twisted, bent awn 5"-8" long, the other spikelet staminate, awnless, consisting of 4 scales   Stamens 3, anthers yellow, orange, or brownish

Dry or moist soil   August to September.

Maine and Ontario to the Rocky Mountains, south to Florida and Texas

INDIAN GRASS

Indian Grass can hardly be passed unnoticed by the wayfaring man, even though he knows little of the herbage of the fields Tall stems leaves a foot in length, and panicles painted in colours of autumn are too striking to be ignored, although they are "nothing but grass"

Blooming in late summer, when the earlier grasses have faded, the long, hairy panicles of Indian Grass are not uncommon in dry fields and in dry places by the waysides   The stems and leaves are often deeply coloured, while the fertile spikelets are brilliant in chestnut-tinted scales and yellow anthers   The soft, densely flowered panicles are rather narrow, and the perfect spikelets are awned, but the sterile spikelets are so reduced and altered that they resemble tiny plumes

Sorghastrum nutans

Indian Grass. *Sorghástrum nùtans* (L.) Nash.

Stem 3-8 ft. tall, erect. Ligule 1"-2" long. Leaves 6'-18' long, 2"-8" wide.

Panicle 4'-12' long, dense, branches erect or slightly spreading. Spikelets 1-flowered, in pairs or 3's; 1 spikelet of each group sessile and perfect; sterile spikelets reduced to hairy pedicels; perfect spikelets 3"-4" long, hairy, shining chestnut brown; scales 4; flowering scale bearing a twisted awn 5"-10" long. Stamens 3, anthers yellow.

Dry soil, fields, waysides, and borders of woods. August to October.

Ontario to Manitoba, south to Florida, Texas, and Arizona.

LARGE CRAB-GRASS, SMALL CRAB-GRASS, AND PURPLE FINGER-GRASS

The two Crab-grasses, large and small, are among the many weeds that have obtained a foothold in America by smuggling their seeds through the port of entry with those of more important plants. Many of the most common weeds — how many can hardly be known — are those that have emigrated with the white man and have tirelessly followed his footsteps through the New World. Such unwelcome foreigners usually take the highways of civilization for their own, and remaining near waysides and in cultivated lands keep the agriculturist forever busy "plucking up the naughty weeds."

56

LARGE CRAB-GRASS *D.   n*   *L*

Large Crab-grass is a weed only when it is out of place, as it so frequently is in this country. In some localities the stems yield a valued pasturage, and in southwestern Europe this grass is cultivated for its seeds, which are used in porridge.

Small Crab-grass (*Digitària humifùsa*) and Large Crab-grass bloom in midsummer and later and are very similar in appearance, differing chiefly in size and in the number of spikes. Small Crab-grass is usually less common, and the second scale of each spikelet is much longer than is the second scale of the larger species. In many a dooryard and near many a garden Large Crab-grass is the most noticeable growth of August and September, when the dark green stems spread over the ground and lift their narrow, deeply coloured spikes which, from the summits of the stems, spread widely, like the rays of an umbel, or like the open fingers of a hand. Near the Large Crab-grass we often notice the contrast of great spreading panicles of Old Witch-grass (*Pànicum capillàre*) which raises its blossoming-heads like shower-fountains of green, and soon rigidly extends the slender branches until the panicles are sometimes two feet across.

Purple Finger-grass, or Slender Finger-grass, (*Digitària filifórmis*) is a native species and therefore is not so often found near dwelling houses. It is an exceedingly

Large Crab-grass
Digitaria sanguinalis

delicate grass in stem, leaf, and inflorescence, the filiform spikes carrying out the slender character of the plant by remaining nearly erect instead of spreading from the stem. This grass is often so slender as to be little noticed, but in early autumn we may frequently find it blooming in sandy fields and by dry roadsides where the Wild Oat-grass still retains dry and faded panicles of springtime. The glistening stems of Purple Finger-grass are often beautifully tinged in rose and purple, colours which, though not looked for among the grasses, are theirs during many months of the year.

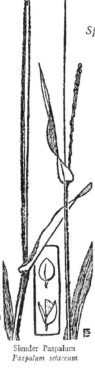

Large Crab-grass. *Digitària sanguinàlis* (L.) Scop.

Annual. Naturalized from Europe.

Stem 1-3 ft. in length, much branched, erect or spreading and rooting at lower joints. Sheaths rather loose, smooth or hairy. Ligule short. Leaves 2'-10' long, 2''-6'' wide, often hairy, rough on edges.

Spikes 4-15, often deep reddish purple, 2'-7' long, narrow, 1-sided, spreading from summit of stem. Spikelets 1-flowered, lanceolate, acute, 1''-1½'' long, in pairs or 3's on one side of the flat rachis. Scales 4; lowest scale minute; 2nd scale about half as long as spikelet. Stamens 3, anthers small. Stigmas lavender.

Cultivated grounds and waste places. July to September.

Throughout North America, except in the extreme north.

SLENDER PASPALUM AND FIELD PASPALUM

Paspalums are characteristic grasses of the Southern States, and in warm countries take the place of the abundant Fescues and Bent-grasses of Northern fields. There are many species, some tall and stout, and others low and spreading, rooting at the joints and carpeting the ground with a dense growth.

Two species only are common in the North, and these, the Slender Paspalum and the Field Paspalum, are low-growing grasses which do

Slender Paspalum
Paspalum setaceum

60

not bloom until midsummer and later. The plump flowers are borne in very narrow, one-sided spikes which even before blooming seem beaded with ripened seed.

Walk through a dry field in late July, and where the earlier grasses have matured and faded green spikes of Slender Paspalum are seen just peeping from their enclosing sheaths. The terminal spike is borne on a slender stem which at length rises many inches above the short upper leaf, while later, other spikes on shorter stems usually protrude from the same sheath. The blossoms of this species are slightly smaller than are those of the Field Paspalum (*Páspalum laève*) which blooms at the same season in moister locations. The two species are distinguished not alone by the more hairy leaves of Slender Paspalum, but also by the fact that the Field Paspalum bears two to five spikes where the other species commonly bears but one.

Slender Paspalum. *Páspalum setàceum* Michx.

Perennial

Stem 1-2 ft. tall, slender, erect or spreading. Ligule short. Leaves and sheaths hairy, leaves 3'-7' long, 1"-3" wide, flat

Spike 2'-4' long, 1-sided, very slender, usually solitary on a long peduncle, additional solitary spikes on shorter peduncles from the sheaths of upper leaves, spikelets 1-flowered, green, about ¾" long, round on outer surface, flat on inner surface. Scales 3. Stamens 3.

Dry fields. July to September

Massachusetts to Nebraska, south to Florida and Texas.

THE PANIC-GRASSES

Panic-grasses are bewildering in their profusion and their variety. No other genus of the grass family offers such a number of species in the Eastern States. Abundant by waysides, in old fields, and on river banks, Panic-grasses are equally common on sandy soils near the coast. Diverse in form, low species, often less than a foot in height, are like miniature bushes, slender ones are lost amid the surrounding taller growth, broad-leaved Panic-grasses, shoulder-high, form dense green thickets by our roadsides, stout species, burned by a hot sun to purple and copper colour, grow in clumps on the beaches, and with long rootstocks bind the wind-blown sands, and a more delicate Panic-grass bearing great flowering-heads of long, hair-like branches is a common tumbleweed in many states. In some species the pyramidal flowering-

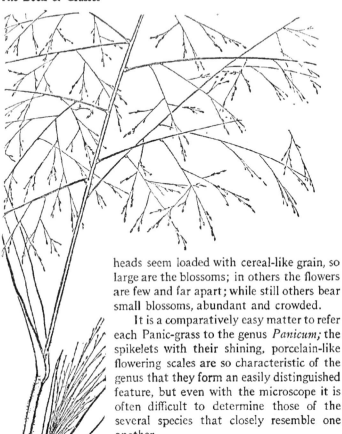

Old Witch-grass
Panicum capillare

heads seem loaded with cereal-like grain, so large are the blossoms; in others the flowers are few and far apart; while still others bear small blossoms, abundant and crowded.

It is a comparatively easy matter to refer each Panic-grass to the genus *Panicum;* the spikelets with their shining, porcelain-like flowering scales are so characteristic of the genus that they form an easily distinguished feature, but even with the microscope it is often difficult to determine those of the several species that closely resemble one another.

Cockspur Grass, formerly included in this genus, which it closely resembles, save in its awn-pointed scales, is common in cultivated lands, where its coarse, erect panicles blossom soon after midsummer. The plant varies greatly, sometimes clothing the flowering-heads in long awns, and again appearing practically awnless. In rich soil the plants are often six feet tall, but in low grounds near thickets and brooks this grass sometimes blooms when it is less than six inches in height. Old

OLD WITCH-GRASS — *Panicum*

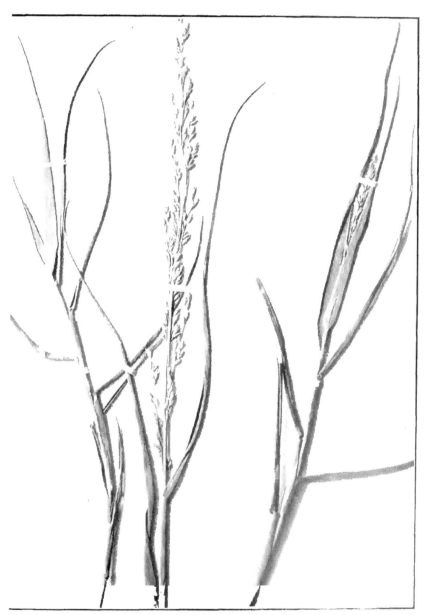

BITTER PANIC-GRASS.

CKSPUR GRASS (*Echinochloa crusgalli*). One-third natural size. A[...] [...] resemble
E. Walteri; the latter may be recognized by its rough sheaths.

A RANK GROWTH OF SALT MARSH COCKSPUR GRASS (*Echinochloa Walteri*) in moist soil near a thicket. Spikelets natural size

Witch-grass (*Pánicum capillàre*), is a beautiful weed of dry soil and is easily recognized, since, even to the most superficial observer, it is unlike any other grass of late summer   The large, yet delicate, flowering-heads, composed of innumerable fine branches, are soft and silky when they first break from the enclosing sheaths, later, when the flowers bloom and the seeds ripen, the panicles are widely open and stiff, and, soon broken by the wind, are blown as tumble-weeds across the fields to scatter their seeds and give them to the care of another year

Small members of the genus are common by roadsides and in fields, where in short flowering-heads of green and purple spikelets Panic-grasses bloom from spring until autumn   Among those most commonly found are the Forked Panic-grass (*Pánicum dichótomum*) bearing short, spreading leaves on slender, wiry stems which support scantily flowered panicles, Starved Panic-grass (*Pánicum depauperàtum*) recognized by its narrow, erect leaves which often equal the stem in height, the several species of Hairy Panic-grass clothed in short, white hairs, Scribner's Panic-grass, with short, coarse leaves, rough on the margins, and borne on hairy sheaths, the panicles composed of large, plump flowers, Round-fruited Panic-grass (*Pánicum sphaerocárpon*) bearing many flowered, purple panicles, hairy-margined sheaths, and broad, rough-edged leaves, Small-fruited Panic-grass (*Pánicum microcárpon*) with longer, oblong panicles of many tiny flowers, and Porter's Panic-grass (*Pánicum Bóscii*) bearing a few large spikelets, the nodes of the stems barbed with soft hairs, and the sheaths and broad leaves clothed in soft pubescence

The larger Panic-grasses do not begin to bloom until midsummer or later   Many of these are common in damp soil   The Large-fruited Panic-grass (*Pánicum latifòlium*) bears pyramidal panicles of large, seed-like spikelets, in Hispid Panic-grass, which, like the preceding species, borders with leafy stems our wayside thickets, the spikelets are more oblong and the sheaths are roughened with short, stiff hairs   Spreading Panic-grass (*Pánicum dichotomiflòrum*) branches abundantly, spreading over the surface of the ground, and the smooth, stout stems and flattened sheaths are surmounted by flowering-heads which are often a foot or more in length

Sea-beach Panic-grass (*Pánicum amaroìdes*) and Tall Smooth Panic-grass (*Pánicum virgàtum*) are characteristic plants of the

71

Scribner's Panic-grass
Panicum Scribnerianum

HIS PID PANIC

Panicum clandestinum

beaches. The former species is distinguished by narrow panicles of erect branches, the latter species (which is also found in sandy soil inland) by large, widely opened flowering-heads brilliantly painted with vivid-coloured anthers and stigmas.

Nor is the above a complete list of the Panic-grasses that bloom in Eastern States: in some localities nearly a score of species may be gathered within the radius of a mile. Brilliant tints are given to many, purple and reddish brown colour the green spikelets; orange and terra-cotta tinge the anthers of some, purple the anthers of others. Often in the smaller species a few leaves are dyed in crimson, and in the majority of the Panic-grasses the feathery stigmas are of deep purple.

Cockspur Grass. Barnyard Grass.
Echinóchloa crusgálli (L.) Beauv.

Annual. Naturalized from Europe.

Stem 1-6 ft. tall, coarse, erect, branching. Sheaths usually smooth, flattened. Ligule wanting. Leaves 6'-24' long, 3''-12'' wide, rough margined.

Panicle 3'-12' long, coarse, branches erect or spreading, densely flowered on lower side. Spikelets ovate, 1-flowered, 1''-1½'' long. Scales 4; 1st scale minute, 2nd and 3d scales rough, about equal, 3d scale awned or awnless; flowering scale shining, enclosing a palet of similar texture. Stamens 3. A variable species. Cultivated grounds, waste places, and by ditches. August to September. Throughout North America, except in the extreme north.

Cockspur Grass
Echinochloa crusgalli

73

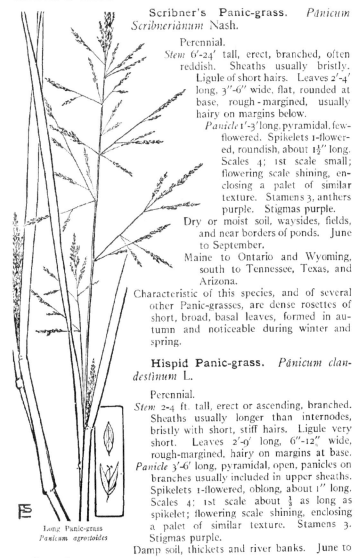

Long Panic-grass
Panicum agrostoides

Scribner's Panic-grass. *Pánicum Scribneriánum* Nash.

Perennial.

Stem 6'-24' tall, erect, branched, often reddish. Sheaths usually bristly. Ligule of short hairs. Leaves 2'-4' long, 3''-6'' wide, flat, rounded at base, rough - margined, usually hairy on margins below.

Panicle 1'-3' long, pyramidal, few-flowered. Spikelets 1-flowered, roundish, about 1½'' long. Scales 4; 1st scale small; flowering scale shining, enclosing a palet of similar texture. Stamens 3, anthers purple. Stigmas purple.

Dry or moist soil, waysides, fields, and near borders of ponds. June to September.

Maine to Ontario and Wyoming, south to Tennessee, Texas, and Arizona.

Characteristic of this species, and of several other Panic-grasses, are dense rosettes of short, broad, basal leaves, formed in autumn and noticeable during winter and spring.

Hispid Panic-grass. *Pánicum clandestínum* L.

Perennial.

Stem 2-4 ft. tall, erect or ascending, branched. Sheaths usually longer than internodes, bristly with short, stiff hairs. Ligule very short. Leaves 2'-9' long, 6''-12'' wide, rough-margined, hairy on margins at base.

Panicle 3'-6' long, pyramidal, open, panicles on branches usually included in upper sheaths. Spikelets 1-flowered, oblong, about 1'' long. Scales 4; 1st scale about ⅓ as long as spikelet; flowering scale shining, enclosing a palet of similar texture. Stamens 3. Stigmas purple.

Damp soil, thickets and river banks. June to September.

Quebec to Michigan, south to Georgia and Texas.

THE DESERTED GARDEN where grow Foxtail Grass, Crabgrass, Cockspur Grass, Old Witchgrass and Snake Grass

GREEN FOXTAIL, YELLOW FOXTAIL, BRISTLY FOX-TAIL, AND ITALIAN MILLET

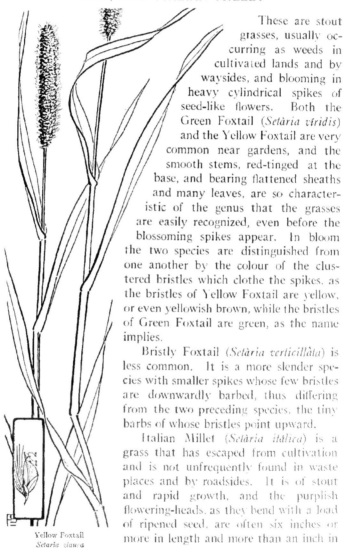

These are stout grasses, usually oc-curring as weeds in cultivated lands and by waysides, and blooming in heavy cylindrical spikes of seed-like flowers. Both the Green Foxtail (*Setària víridis*) and the Yellow Foxtail are very common near gardens, and the smooth stems, red-tinged at the base, and bearing flattened sheaths and many leaves, are so character-istic of the genus that the grasses are easily recognized, even before the blossoming spikes appear. In bloom the two species are distinguished from one another by the colour of the clus-tered bristles which clothe the spikes, as the bristles of Yellow Foxtail are yellow, or even yellowish brown, while the bristles of Green Foxtail are green, as the name implies.

Bristly Foxtail (*Setària verticillàta*) is less common. It is a more slender spe-cies with smaller spikes whose few bristles are downwardly barbed, thus differing from the two preceding species, the tiny barbs of whose bristles point upward.

Italian Millet (*Setària itàlica*) is a grass that has escaped from cultivation and is not unfrequently found in waste places and by roadsides. It is of stout and rapid growth, and the purplish flowering-heads, as they bend with a load of ripened seed, are often six inches or more in length and more than an inch in

Yellow Foxtail
Setaria glauca

thickness Millets were among the most ancient of cultivated
grains, being planted long ago in China each spring by princes of
the royal house And in lake dwellings of the Stone Age the
grain has been found in such quantities that it must be assumed
to have yielded the chief bread supply of prehistoric men

Yellow Foxtail. Pigeon-grass. *Setaria glaúca* (L)
Beauv.

Annual Naturalized from Europe
Stem 1-4 ft tall, smooth, branched, erect Lower sheaths loose and
flattened Ligule a ring of short hairs Leaves 2'-12' long, 2"-5"
wide, somewhat hairy at base
Spike (spike-like panicle) 1'-4' long, cylindrical, densely flowered, clothed
in tawny yellow bristles Spikelets 1-flowered, about 1½" long, surrounded by a cluster of 5-10 upwardly barbed bristles which rise from
below the base of each spikelet and exceed the spikelet in length
Scales 4, outer scales unequal, 3d scale sometimes enclosing a palet
and staminate flower, flowering scale of perfect flower wrinkled, thick,
and very convex Stamens 3, purple Stigmas purple
Cultivated ground and waste places July to September
Throughout North America, except in the extreme north

BUR-GRASS

Nature has decreed that the gatherers of her harvests shall be
disseminators of the plants they use But those weeds which
are tramps of the wayside, like Spanish needles, burdock, and Bur-
grass, having nothing of value to commend their transportation
to new fields, have developed an insistent scheme for pushing new
generations out into the world, and bind burdens upon all passers-
by Most appropriately do botanists comment upon Bur-grass as
"a vile and annoying weed," since it is one that causes trouble
from the Atlantic to the Pacific No one who has walked along
railways, to find them Elysian fields in their variety of flora, can
forget climbing sandy embankments through a hindering growth
of this plant

Bur-grass is more abundant in the Southern than in the Northern states It is low and spreading, sometimes carpeting the
ground on waste land and near sandy shores, and the flowering
spikes, composed of numerous, spiny burs, present a peculiar appearance, leading one to doubt if this can be grass at all A reddish tinge is often noticeable in the flat sheaths as well as in the

GREEN FOXTAIL (*Setaria viridis*)   Plate 30 Original.

a   b   c   d

stems, and in the burs, which are really involucres enclosing the
spikelets. Before the blossoming-head breaks from the sheath
each involucre is short
and cup-like, sur-
mounted by broad
green bristles, but at
maturity these bristles
are grown together into a
hard bur which encloses the
seeds and is beset with spines
of needle-like sharpness. Later
in the season the burs readily
become detached and, adhering to
passing objects, are carried long
distances until they fall on new soil
where the seeds establish new colonies
of this troublesome grass.

**Bur-grass. Sand-grass. Devil-
burs. Hedgehog-grass.** *Cénchrus caroliniànus* Walt.

Annual.

Stem 6'-24' in length, much branched, erect or
spreading. Sheaths loose, smooth, flattened. Ligule a ring of short hairs.
Leaves 2'-6' long, 2"-4" wide.

Spike 1'-3' long, composed of 6-20 round,
spiny burs enclosing the spikelets; burs
more or less downy, sometimes reddish;
spines very rigid at maturity. Spikelets
2-flowered, about 3" long. Scales 4, thin.
Stamens 3.

Sandy soil. July to September.

Maine and Ontario to South Dakota, south to
Florida, Texas, and southern California.

INDIAN RICE

"And I will cut a reed by yonder spring
And make the wood-gods jealous."

Many who are but superficially familiar with the low herbage of the fields
hesitate to name as grass such large

Bur-grass.
Cénchrus caroliniànus.

83

plants as Indian Rice, and attempt to solve the question by calling them reeds. But reed or grass, it is the same, and grass-like characteristics are constant whether measured by inches or disguised by a gigantic growth.

In shallow water and on muddy shores the Indian Rice grows, a tall, stout grass whose long flowering heads seem like a combination of flowers from two dissimilar plants; the upper, fruit-bearing portion of the panicle consisting of narrow, erect branches with long-awned flowers, while below them awnless, staminate flowers droop from branches that are widely spreading. The dark seeds are half an inch or more in length, and where the grass grows by lakes in Minnesota and the Northwest the Indians paddle their canoes among thickets of Indian Rice and beat off the grain, gathering it as a cherished article of food, while in the water hungry fishes eagerly eat the scattered seed.

On the Jersey marshes, and south-ward by tidal waters of the Middle States, multitudes of bobolinks in sober dress stop, during their journey toward warmth and sunshine, and find bountiful fare spread for them on ripening panicles of this grass. The bobolink's flood of melody poured over June fields is lost in autumn, and name as well as plum-age is changed; ricebird, or reedbird, is the sobriquet under which he travels. These birds are, alas! a favourite target for fall sportsmen, and it may be re-

Indian Rice
Zizania palustris

84

WILD RICE (*Zizania palustris*) growing in the swampy border of a stream

a   b

WILD GRASSES OF JAPAN

membered that the charm bag of Brer Rabbit carried "one ricebud bill."

Indian Rice. Wild Rice. Reeds. *Zizània palústris* L.

Stem 3-10 ft. tall, smooth, stout, erect. Sheaths loose. Ligule about 3" long. Leaves 1-3 ft. long, 3"-16" wide.

Panicle 1-2 ft. long, pyramidal. Spikelets 1-flowered. Scales 2. The upper portion of panicle consists of erect branches bearing narrow pistillate spikelets 4"-12" long; outer scale bearing a rough awn 1'-2' long; the lower portion of panicle consists of widely spreading branches bearing staminate, awnless spikelets 3"-6" long. Stamens 6. Grain about 6" long. Swamps and borders of streams. June to October.

New Brunswick to Manitoba, south to Florida and Texas.

RICE CUT-GRASS AND WHITE-GRASS

Brook borders offer so much of beauty in their flora that the less noticeable leaves and blossoms of such waterside plants as Rice Cut-grass and White-grass often serve only as a background, intensifying the brilliancy of cardinal-flowers and emphasizing the deep blue of gentians. But if one attempts to walk through a tangle of these grasses, in order to reach some flower that grows between the rocks at the water's edge, rough leaves and sheaths clothed in minute, hooked prickles delay progress, and if the hand is used in pushing the grasses

Rice Cut-grass
Leersia oryzoides

89

aside it suffers sorely. Indeed, it is unpleasant to pick even one piece of Rice Cut-grass for analysis, so determinedly do the leaves catch on hands and clothing.

Both White-grass (*Leérsia virginica*) and Rice Cut-grass bloom in late summer in wet places, where the stems, branching abundantly, bear panicles of green or whitish-green blossoms Rice Cut-grass is the stouter species and bears larger panicles, while enclosed in the lower sheaths cleistogamous blossoms may often be found. In White-grass the few branches of the panicles spread stiffly and bear comparatively few spikelets, while the leaves are shorter, broader, and less rough than are those of the larger species

It is said that the leaves of certain species of the genus are sensitive in the same manner as are the leaves of the sensitive plant One species, known as Catch-fly Grass (*Leérsia lenticulàris*) bears wide spikelets armed with strong bristles. Of this grass and its blossoms, Pursh, an early botanist, writes "Found on islands of Roanoke River, N C, and observed it catching flies in same manner as *Dionǣa muscipula* (Venus's Fly-trap) " The scales certainly look as if they might close like steel-traps and imprison insects, but as the most modern text-books say nothing of this habit his record may remain as an "evidence of things not seen" by less fortunate botanists of later times

Rice Cut-grass. Cut-grass. *Leérsia oryzoìdes* (L) Sw.

Perennial

Stem 2-4 ft tall, much branched, erect or spreading. Ligule very short. Sheaths and leaves rough, clothed in minute, downward-pointing, hooked prickles Leaves 4'-10' long, 2''-5'' wide.

Panicle 5'-10' long, branches spreading, not numerous Spikelets 1-flowered, flattened, light green, 2''-2½'' long Scales 2, nearly equal in length, downy, outer scale rough on keel and margins, inner scale rough on keel Stamens 3, anthers pale yellow.

Marshes and along streams July to September

Nova Scotia to Ontario, south to Florida and Texas

REED CANARY-GRASS

"Thou should'st have gathered reeds from a green stream "

A broad ribbon of dull rose often borders the winding streams and brooks of June when, for a few weeks, the Reed Canary-grass blooms. In spikelets and lightly poised anthers this grass offers

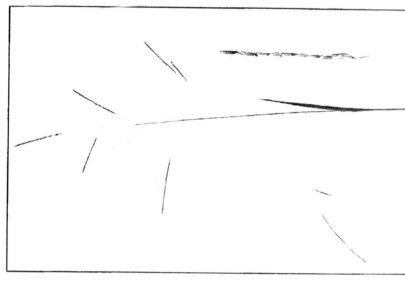

WHITE-GRASS (*Leersia virginica*). Natural size. Fruit enlarged by two

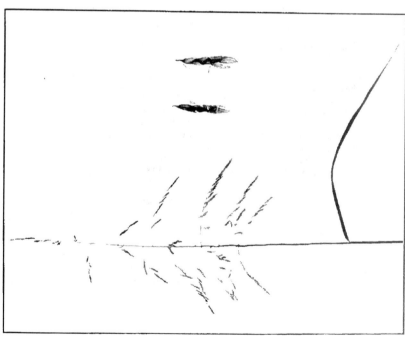

RICE CUT-GRASS (*Leersia oryzoides*). One-half natural size. Spikelets enlarged by two

most attractive colouring, and even before flowering the dense growth of blue-green leaves is noticed in marked contrast to the lighter colour of the marsh.

A stout perennial with creeping rootstocks, this grass grows most luxuriantly in wet meadows and in shallow water, where the profusion of

Enlarged spikelet of
Phalaris arundinacea

leaves, which are always darker in colour than the smooth and shining stems, forms a mass of verdure shoulder-high. In bloom the short branches of the panicles spread from the stem, but they are soon drawn closely to it again as the flowers fade, and in the ripening head the fungus commonly known as ergot often appears as black spurs issuing from between the scales.

The true Canary-grass (*Phalaris canariénsis*), cultivated in

Phalaris arundinacea

Europe for the seeds, which have been used as a cereal as well as for bird food, has been introduced in this country and may now be found in waste places The short, thick spikes are about an inch in length and are strikingly marked by their white and green scales.

Under the names of Ribbon-grass, Lady's Ribbons, Gardeners' Garters, Painted-grass, and French-grass, a variety of the Reed Canary was planted in the gardens of earlier days Gerard, in his "Herball," describes the leaves of this plant as fashioned "like to laces or ribbons woven of white or greene silke, very beautiful and fair to behold," and these striped leaves with their rare, silver-like lustre, are occasionally found by our waysides where the grass has escaped from cultivation

Reed Canary-grass *Phálaris arundinàcea* L.

Stem 2-5 ft tall, stout, erect   Sheaths smooth   Ligule 1″-2″ long
  Leaves 4′-12′ long, 3″-9″ wide, roughish, flat
Panicle 3′-8′ long, densely flowered, open in flower, contracted before and
  after blossoming   Spikelets 1-flowered, 2½″-3″ long, green strongly
  tinged with rose-purple   Scales 5, outer scales rough, about equal,
  3d and 4th scales reduced to hairy rudiments, flowering scale hairy
  Stamens 3, anthers yellow or lavender
Moist ground and shallow water   June to August
Nova Scotia to British Columbia, south to Maryland, Tennessee, and
  Arizona

SWEET VERNAL-GRASS

"Two gentle shepherds, and their sister-wives,
With thee, ANTHOXA! lead ambrosial lives "

The student is fortunate who begins the analysis of grasses with the Sweet Vernal, as did Darwin, who wrote of it   "I have just made out my first grass, hurrah! hurrah! I must confess that fortune favours the bold, for, as good luck would have it, it was the easy *Anthoxánthum odorátum*, nevertheless it is a great discovery, I never expected to make out a grass in my life, so hurrah! It has done my stomach surprising good "

Sweet Vernal is the first grass to attract one in early spring as in April it pushes up its compact, spike-like panicles to expand them soon with the open blossoms, whose large, violet anthers, as in many wind-fertilized plants, furnish the colour that is lacking in the tiny flower

94

REED CANARY GRA . *Phalaris arundinacea* . One thir natural s e) . . . rink

Although many of our common grasses possess a faint and agreeable odour in blossoming time, as well as in the hayfield, Sweet Vernal-grass and Vanilla Grass are our only strongly fragrant species, and after a May shower the sweetness of country air is due, not to the more noticeable blossoms of spring, but to the countless spikes of Sweet Vernal, so abundant by waysides and in upland meadows. The resinous principle, coumarin, to which the fragrance is due, is similar in odour to benzoin, and is found in a number of other plants; one of these, the Blue Melilot, is, in Switzerland, mixed with cheese to which the plant imparts its peculiar odour, and it has been said that in Italy water distilled from Sweet Vernal-grass has been used as a perfume.

The slender, satiny stems of this grass are of beautiful texture and, with those of June Grass, have been used in the weaving of imitation Leghorn hats, as well as in basketry.

Sweet Vernal-grass. *Anthoxánthum odoràtum* L.

Perennial. Naturalized from Europe.
Stem 1-2 ft. tall, slender, erect. Ligule 1″-2″ long. Leaves 1′-6′ long, 1″ 3″ wide, flat, smooth or sparingly downy.
Spike-like Panicle 1′-4′ long, green or brownish. Spikelets 1-flowered, narrow, 3″-4″ long. Scales 5; outer scales very unequal, often downy; 3d and 4th scales hairy and bearing dorsal awns; awn of 4th scale bent and twisted, more than twice the length of the scale. Stamens 2, anthers violet. Stigmas white. Plant very fragrant in drying.
Waysides, meadows, and pastures. April to July.
Throughout nearly the whole of North America.

VANILLA GRASS

"And because, the Breath of Flowers, is farre Sweeter in the Aire, (where it comes and Goes, like the Warbling of Musick) then in the hand, therefore nothing is more fit for that delight, then to know, what be the Flowers, and Plants, that doe best perfume the Aire."

Sweet Vernal-grass
Anthoxanthum odoratum

97

In all the fragrance of outdoor life the grasses bear their part in diffusing that most subtle essence which is as untraceable as the

odour of spring. Vanilla Grass is one of the most fragrant grasses, and its perfume in the air suggests that of delicate orchids hidden among sedges and rushes by the brookside.

Vanilla Grass is found in abundance only in the more Northern States and in Canada, but as far south as New Jersey it is occasionally found in damp soil. In bloom the short panicles are ornamented in chestnut-brown and purple, but it is the long-leaved sterile shoots which have gained for the plant a wide recognition far from its native meadows. These leaves are very fragrant in drying, and are gathered by Indians who weave them into many varieties of baskets and placques. It is even said that European peasants attach to the plant some mysterious power of inducing sleep, and that bunches of the leaves are sold in the north of Europe to be suspended in sleeping-rooms.

When, in olden time, the custom still lingered of scattering sweet-scented grasses before the churches of northern Europe on Saints'-days and holy festivals, this plant received the name of Holy Grass and was used with the less strongly scented Sweet Vernal-grass to make fragrant the pathways leading to shrines of the saints.

Fragrant grasses were at one time much lauded by agriculturists, who thought that sweet grasses should improve the flavour of butter and milk, but alas for theories! it was proved that Vanilla Grass and Sweet Vernal-grass were not only valueless, but were even disliked by cattle.

VanillaGrass. Holy Grass. Seneca Grass.
Hieróchloë odoràta (L.) Wahlenb.

Perennial, from creeping rootstocks.
Stem 1-2½ ft. tall, slender, erect. Ligule 1"-2" long.
Stem leaves very short, basal leaves, and those of sterile shoots, long and narrow.

Vanilla Grass
Hierochloë odorata

98

Panicle 2'-4' long, pyramidal, open, somewhat 1-sided. Spikelets 3-flowered, 2"-3" long, chestnut-brown or purplish; upper flower perfect; lower flower staminate. Scales 5; outer scales nearly equal, smooth; 3d and 4th scales hairy; flowering scale hairy at apex. Stamens 2 or 3. A very fragrant grass.

Moist grounds. May to July.

Newfoundland to Alaska, south to New Jersey and Colorado.

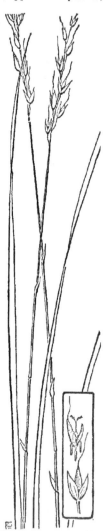

WHITE-GRAINED MOUNTAIN RICE, BLACK-GRAINED MOUNTAIN RICE, SLENDER MOUNTAIN RICE

"The grass flames up on the hillsides like a spring fire . . . not yellow but green is the colour of its flame; the symbol of perpetual youth."

The early flowers of Northern woods, arbutus, yellow adder's tongue, and dicentra, have long been honoured as heralds of spring, while the lesser blossoms of shade-loving early grasses and sedges have been passed unnoted. Yet, the grasses, too, brave inclement winds and frosts, and without them the forest floors would seem bare indeed.

White-grained Mountain Rice is one of the earliest of the woodland grasses, and the common name refers to the large seed of which it is said that a white flour has been made. The tufts of long basal leaves remain green during the winter, and in early spring the slender stems rise (often purple-tinged at the base) bearing short, narrow panicles of a few pale-coloured flowers.

Slender Mountain Rice (*Oryzópsis púngens*) is another grass of early spring, and, although it is usually less common than the other species, it may be looked for near open woods and on dry and rocky soil. The

White-grained Mountain Rice
Oryzopsis asperifolia

99

short, narrow panicle rises above thread-like leaves, and the whitish spikelets are occasionally tinged with purple.

Black-grained Mountain Rice (*Oryzópsis racemòsa*) is a later, broad-leaved species found in woods and on ledges, where it blooms in midsummer. The plant is larger than White-grained Mountain Rice and usually bears a longer panicle of darker flowers; nor are the stem-leaves minute, as are those of that earlier species.

White - grained Mountain Rice. Winter-grass. *Oryzópsis asperifòlia* Michx.

Perennial.

Stem 8'-20' tall, slender, erect. Ligule very short. Stem leaves very short, basal leaves long and narrow, 2"-4" wide.

Panicle 2'-4' long, few-flowered, narrow and contracted. Spikelets 1-flowered, broad, 3"-4" long. Scales 3; outer scales slightly unequal; flowering scale whitish, sparingly downy, bearing a terminal awn 4"-5" long. Stamens 3. The leaves remain green during winter.

Woodlands. April to June.

Newfoundland to British Columbia, south to New Jersey, Pennsylvania, and Minnesota; also in the Rocky Mountains to New Mexico.

BLACK OAT-GRASS

Through some selective process of Nature her longest-awned grasses are most frequently found on dry and rocky soil. The Aristidas, and the different genera known under the name of Oat-grasses, often grow near dry woods, in that interesting borderland where wild black-

Black Oat-grass
Stipa avenacea.

berries and strawberries crowd beneath white birches and spire-like cedars   Bottle-brush Grass retreats farther into the woods, and Black Oat-grass is also found within their shelter, usually on the southern slope of some open woodland where the sunlight penetrates the leafy shade

The greater number of the species of this genus (*Stipa*) are found west of the Mississippi, and form the Bunch-grasses, Feather-grasses, and Needle-grasses of the plains

Black Oat-grass is common eastward in early summer, when the loose, few-flowered panicles rise above the tightly rolled and thread-like leaves.   The ripened flowering-heads often remain on this grass until autumn, and the long awns, bent near the middle and twisted below, spread widely from the scales.   These twisted awns uncoil during damp weather but coil tightly again when the sun shines, and from this habit the Stipas, in older days, were known as "weather grasses"

A foreign species, with silky-feathered awns nearly a foot in length, has been cultivated for its beauty, and a species of southern Europe is an important article of commerce, baskets, ropes, and paper being made from the tough leaves

Black Oat-grass.   *Stipa avenàcea* L

Perennial

Stem 1-3 ft tall, slender, erect   Ligule about 1″ long   Leaves in-volute, thread-like, stem leaves 3′-5′ long, basal leaves longer

Panicle 4′-8′ long, few-flowered, open, branches slender.   Spikelets 1-flowered, narrow, 4″-5″ long   Scales 3, outer scales narrow, nearly equal, acute, flowering scale blackish, hairy at base and bearing a bent, loosely twisted, terminal awn about 2′ long   Stamens 3

Dry, open woods   May to July

Southern New England to Ontario and Wisconsin, south to Florida and Mississippi.

POVERTY GRASS, SLENDER ARISTIDA, PURPLISH ARISTIDA, AND SEA-BEACH ARISTIDA

The height of the grass season continues through all the summer months, and even in September the student will find fresh grasses by the wayside   These are not always giant species like the Reed, which uses the entire season in maturing its growth, several of the smaller grasses of wiry stem and narrow leaf have, in the long evolution of the grass family, found it more advantageous to wait until the pressure of midsummer was past before giving their seeds to the care of Nature.

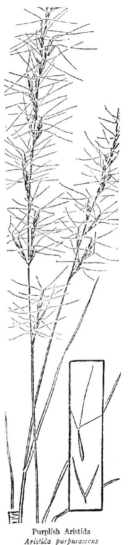

Among other grasses of late summer the Aristidas are common in dry soil throughout the country. The English name of Three-awned Grass is descriptive of a peculiarity of the genus, as each flowering scale bears triple awns. In Poverty Grass and Slender Aristida the outer awns of the flowering scale are shorter than the middle awn and are upright, while the long middle awn spreads stiffly at right angles to the spike. When the spikelets are comparatively few, as in the species mentioned above, these horizontally spreading awns are so characteristic that from them alone the grasses may easily be recognized.

Poverty Grass (*Aristida dichótoma*) is the smallest of the eastern Aristidas and bears but short awns. Slender Aristida has a slightly larger flowering-head whose horizontal awns are frequently one half inch in length. The panicles of Purplish Aristida are long and very bristly; the outer awns of each flowering scale nearly equal the horizontal middle awn in length, and

Slender Aristida
Aristida gracilis

Purplish Aristida
Aristida purpurascens

ARUACH ARETIDA

spread slightly from the spike-like purple panicle; the plant is usually larger than either of the preceding species and bears longer leaves   The awns of Sea-beach Aristida (*Arístida tuberculòsa*) are of nearly equal length and are united at their base for one quarter of an inch or more   The panicles of this grass are few-flowered, and the awns are widely spreading or even reflexed

The several species of eastern Aristidas are locally known as Poverty Grasses, from their appearing most frequently on waste land and on soil that is too poor to support a richer vegetation. Many species are common in the West and Southwest, where, among others, is found the well-named Needle-grass, whose triple awns sometimes attain a length of four inches.

Slender Aristida.   *Arístida grácilis* Ell.

Annual

Stem 6'-24' tall, slender, erect, often branched   Ligule very short   Leaves bristle-like, 1'-4' long, hardly 1" wide

Spike-like Panicle 3'-7' long, slender, not densely flowered.   Spikelets 1-flowered, narrow, about 3" long   Scales 3, narrow, outer scales nearly equal, awn-pointed; flowering scale bearing 3 awns, middle awn horizontal, 3"-9" long, lateral awns erect, 1"-4" long   Stamens 3

Dry and sandy soil   August and September

Massachusetts to Nebraska, south to Florida and Texas

Purplish Aristida.   *Arístida purpuráscens* Poir.

Perennial

Stem 1-3 ft tall, slender, erect   Ligule very short   Leaves long and narrow, sometimes involute, 4'-10' long, about 1" wide.

Spike-like Panicle 6'-18' long, slender   Spikelets 1-flowered, 3"-5" long, narrow, purplish   Scales 3, narrow, outer scales unequal, flowering scale bearing 3 awns, middle awn horizontal, about 1' long, lateral awns slightly shorter, erect or spreading   Stamens 3

Dry and sandy soil, and in dry woods   August to October.

Massachusetts to Minnesota, south to Florida and Texas

THE MUHLENBERGIAS

WOOD MUHLENBERGIA, ROCK MUHLENBERGIA, MARSH MUHLEN-
BERGIA, MEADOW MUHLENBERGIA, NIMBLE WILL, AND
LONG-AWNED HAIR-GRASS

With one exception the common species of this genus are unattractive grasses, which, although they add the verdure of their leaves to waysides and to country dooryards, bear but inconspicuous flowering-heads of little beauty. As the Muhlenbergias are native grasses they are found in many localities, and grow from open woods to dry fields and on moist banks of streams. In certain soils the tough, matted root-stocks are but too frequently seen in gardens. The smooth stems, rising in early summer, are very leafy, but as the season advances and the stems lengthen the leaves thereby become more remote, and in maturity the grasses are hard and wiry. All the species bloom in late summer, and throughout the season the fresh plants have a taste peculiar to the genus.

Wood Muhlenbergia (*Muhlenbérgia sylvática*) and Nimble Will (*Muhlenbérgia Schrebèri*) are frequent along the borders of woods, and in rocky places one naturally looks for Rock Muhlenbergia (*Muhlenbérgia sobolífera*). These are slender grasses that are usually much branched and that bear narrow, spike-like panicles of small, green flowers.

Marsh Muhlenbergia (*Muhlenbérgia racemòsa*) grows in wet places and has much stouter and more compact flowering-heads, which sometimes resemble spikes of Timothy.

Meadow Muhlenbergia frequently grows near dwelling houses, where in early summer it offers the contrast of spreading clumps of yellowish green leaves to the darker colour of June Grass and Orchard Grass. Meadow Muhlenbergia is the last of the common dooryard grasses to bloom, and after one has watched the branching stems the season through, and has waited with

Meadow Muhlenbergia
Muhlenbergia mexicana

106

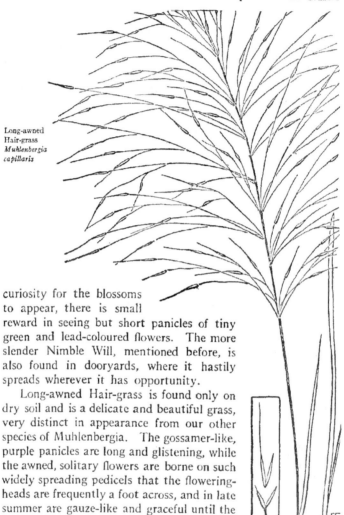

Long-awned
Hair-grass
*Muhlenbergia
capillaris*

curiosity for the blossoms
to appear, there is small
reward in seeing but short panicles of tiny
green and lead-coloured flowers. The more
slender Nimble Will, mentioned before, is
also found in dooryards, where it hastily
spreads wherever it has opportunity.

Long-awned Hair-grass is found only on
dry soil and is a delicate and beautiful grass,
very distinct in appearance from our other
species of Muhlenbergia. The gossamer-like,
purple panicles are long and glistening, while
the awned, solitary flowers are borne on such
widely spreading pedicels that the flowering-
heads are frequently a foot across, and in late
summer are gauze-like and graceful until the
first frost touches them.

Meadow Muhlenbergia. Mexican Dropseed. *Muhlenbérgia mexicàna* (L.) Trin.

Perennial, from creeping rootstocks.

Stem 1½-3½ ft. tall, smooth, wiry, much branched, erect or spreading.

Ligule less than 1″ long.   Leaves 3′-8′ long, 1″-4″ wide, roughish.
Panicle 2′-6′ long, narrow, many-flowered.   Spikelets 1-flowered, 1″-1½″ long, green or bluish purple.   Scales 3; outer scales acute, slightly unequal; flowering scale acute.   Stamens 3.   An exceedingly variable species.

Damp soil and in shaded places.   August and September.

New Brunswick to Ontario, south to North Carolina and Oklahoma.

Long-awned Hair-grass.   *Muhlenbérgia capillàris* (Lam.) Trin.

Perennial.

Stem 1½-3½ ft. tall, slender, erect, not branched.   Ligule 1″-2″ long.   Leaves long and narrow, 6′-15′ long, 1″-2″ wide.

Panicle 6′-18′ long, open, delicate, usually purple, branches hair-like, lower branches 3′-8′ long.   Spikelets 1-flowered, about 2″ long, narrow, on hair-like pedicels.   Scales 3; outer scales slightly unequal, acute; flowering scale bearing a slender, terminal awn 3″-10″ long.   Stamens 3.

Sandy and rocky soil and open woods.   August to October.

Massachusetts to Missouri, Florida, and Texas

LONG-AWNED WOOD-GRASS

The narrow brooks threading their way through woods and swamps are the haunts of many plants whose location makes them the more rare to the ordinary pedestrian. Here, where the bladderwort hangs its tiny yellow sunbonnets far from travelled paths, and the wild calla unfolds pallid against the velvet mud, may be found the Long-awned Wood-grass growing luxuriantly on a dryer bank of the brookside.

The grass is distinctly graceful, and at first glance the slender, nodding panicle might suggest the flowering-head of a Brome-grass, but the form of the leaves separates the Long-awned Wood-grass at once from that genus, while the one-flowered spikelets differ from

Long-awned Wood-grass
Brachyelytrum erectum

108

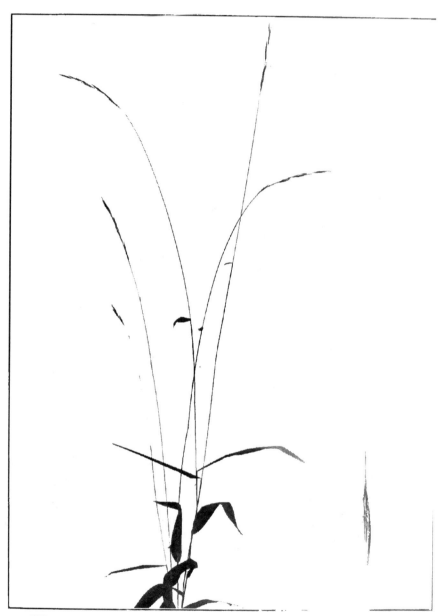

LONG-AWNED WOOD-GRASS.   so and a half

the many-flowered spikelets of the Brome-grasses. If the microscope is used it discloses the thread-like prolongation of the rachilla lying against the palet which closely embraces the narrow seed. The slender stems bear a profusion of spreading leaves, and the whole plant has a slightly unpleasant odour.

Long-awned Wood-grass. *Brachyélytrum erēctum* (Schreb.) Beauv.

Perennial from rootstocks.

Stem 1-4 ft. tall, slender, erect. Nodes and sheaths downy. Ligule less than 1″ long. Leaves 2′-6′ long, 3″-8″ wide, rough above, downy on lower surface, tapering at both ends, flat.

Panicle 2′-7′ long, narrow, slender, few-flowered, branches erect, 1′-4′ long. Spikelets 1-flowered, narrow, 5″-6″ long. Scales 3; outer scales small, unequal; 1st scale very minute; flowering scale 4″-6″ long, 5-nerved, bearing a rough, terminal awn 9″-12″ long. Rachilla prolonged and lying as a slender bristle in the groove of the palet. Stamens 2, anthers and stigmas long. Plant has faint unpleasant odour.

Open woods and moist grounds. June to August.

Newfoundland to Ontario and Minnesota, south to North Carolina and Missouri.

TIMOTHY

"It is full summer now, the heart of June."

In many of the states Timothy is one of the most common of cultivated grasses, and it is, perhaps, the one most generally known and easily recognized. By waysides and in fields its bright green bayonets rise, stiff and rigid, and tipped with cylindrical flowering-heads which bloom at about the same time as Red-top, a grass with which this species is often associated in the fields, though Timothy is noticeable earlier in the season, when the sunlight, touching the tips of the spikelets, seems to gem the spikes with dew-drops. The flowers are densely crowded, and in bloom the

Timothy
Phleum pratense

111

anthers, borne on long filaments, increase the apparent size of the heads by encircling them in filmy lavender.

The stalks, in rich soil, are sometimes five feet tall, with blossoming heads six to ten inches in length, but in dry places the plant is much smaller, and by developing bulbous thickenings of stored-up nutriment at the bases of the stems it is enabled to survive periods of drouth. Leaf-smut occasionally attacks this grass and reduces the leaves to shreds covered with dusty brown spores

A few of the flowering-heads of aftermath develop leaves from the tips of the scales and cover the spikes with tiny green blades, as if to shield the unripened seed from early frosts.

The common name of Timothy preserves that of its earliest cultivator, Timothy Hanson, a Maryland planter of the early days of the eighteenth century. He must have been a genial man, proving himself friendly, or we should have had Hanson Grass instead of Timothy, although the more local name of Herd's Grass need be considered no reproach to the Mr Herd who cultivated it in New Hampshire long ago

Timothy. Herd's Grass. Cat's-tail Grass. *Phlèum pra-*
tènse **L.**

Perennial.

Stem 1-5 ft tall, erect, not branched. Ligule 1″-2″ long. Leaves 3′-12′ long, 2″-4″ wide, flat, rough or nearly smooth.

Spike (*spike-like panicle*) 1′-7′ long, cylindrical, green, densely flowered, 3″-4½″ in diameter. Spikelets 1-flowered, flat, about 1½″ long Scales 3, outer scales compressed, about equal, hairy on keels, abruptly awn-pointed, flowering scale thin and translucent, truncate, much shorter than outer scales Stamens 3, anthers usually lavender. Stigmas white

Meadows, fields, and waysides June to August.

Throughout nearly the whole of North America.

MEADOW FOXTAIL AND MARSH FOXTAIL

The resemblance that Meadow Foxtail bears to Timothy might be confusing were it not that the former, being one of the first grasses to mature its seed, begins to bloom a month before the green spikes of Timothy appear The whole plant is more soft than is Timothy, the leaves shorter and borne on somewhat inflated sheaths, and the spikes are slightly softer, broader, and shorter than are the stiff, rough heads of the later-flowering grass Meadow

Foxtail is exceedingly hardy, thriving on all soils but the driest, and after the early growth of May and June it yields later a luxuriant aftermath.

Common in low meadows and along shallow streams is the Marsh Foxtail, (*Alopecùrus geniculàtus*,) a widely distributed grass of rich, dark-green colour. A more slender plant than Meadow Foxtail, it bears shorter spikes and spikelets, while there is an important difference in the empty scales of the two species, those of Marsh Foxtail being not only much shorter but also quite obtuse. The slender stems, bending and spreading at the base, are sparingly branched, and it will be noticed that the upper leaf is as long as its sheath, as is seldom the case in Meadow Foxtail.

Meadow Foxtail. *Alopecùrus praténsis* L.

Perennial, with short rootstocks. Naturalized from Europe.

Stem 1-3 ft. tall, erect, not branched. Sheaths loose. Ligule very short. Leaves 1'-4' long, 1"-3" wide, flat, rough or nearly smooth.

Spike (spike-like panicle) 1'-4' long, cylindrical, green, densely flowered, 4"-6" in diameter. Spikelets 1-flowered, compressed, 2"-3" long. Scales 3; outer scales acute, equal, united at the base, hairy on the keels; flowering scales nearly as long as empty scales, thin and translucent, obtuse, bearing a slender, dorsal awn about 3" long; palet often lacking. Stamens 3. Stigmas long.

Fields and meadows. May to July.

Labrador to southern New York, Ohio, and Michigan, also in Oregon and California.

THE DROPSEED GRASSES

GAUZE-GRASS, SHEATHED RUSH-GRASS, SMALL RUSH-GRASS, LONG-LEAVED RUSH-GRASS, SAND DROPSEED, AND NORTHERN DROPSEED

The common species of this genus are very dissimilar in appearance: the panicle of Gauge-

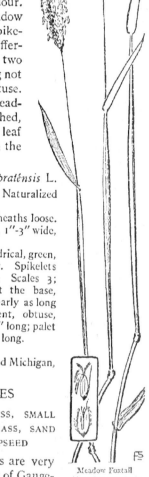

Meadow Foxtail
Alopecùrus praténsis

grass is lacy and open, like that of a miniature Eragrostis; Sheathed Rush-grass and Small Rush-grass bear short, narrow flowering-heads; and the panicles of certain other species show a superficial resemblance to those of the Bent-grasses.

Gauze-grass is so small and delicate that it hides beneath one's feet, and hundreds of the plants may grow unnoticed by the margin of a brook where taller grasses have been gathered. Indeed this grass is seldom seen at all, save by accident, as it were, when the student, down on bended knee, is searching for the smaller flowering plants of wet, sandy soil. Often the Rough Hair-grass grows in similar locations and bends its great ripening panicles over the later-flowering ones of Gauze-grass.

Our other species of these grasses grow in dry or sandy soil and may be looked for in late summer and early autumn, when with a low growth they frequently cover the ground between tufts of the Beard-grasses, or rise among the slender, rose-purple stems of Purple Finger-grass. Small Rush-grass (*Sporóbolus negléctus*) and Sheathed Rush-grass are small and very slender, with short, narrow, leaves, and very short, narrow flowering-heads which, in the latter, bear anthers of deep red in vivid contrast with the white stigmas. Long-leaved Rush-grass (*Sporó-*

Sheathed Rush-grass
Sporobolus vaginaeflorus

Gauze-grass
Sporobolus uniflorus

114

bolus ásper), also, bears a narrow panicle, but the stem is often three feet in height and the leaves are very long

The bluish or lead-coloured panicles of Sand Dropseed (*Sporóbolus cryptándrus*) are open and many-flowered, Northern Dropseed (*Sporóbolus heterólepis*) bears an open, less heavily flowered panicle, and long, thread-like leaves

Sheathed Rush-grass. Southern Poverty-grass. *Sporóbolus vaginaeflórus* (Torr.) Wood

Annual

Stem 6'-20' tall, slender, tufted, erect   Sheaths somewhat inflated
   Ligule very short   Leaves 1'-4' long, 1'' wide or less, upper leaves
   rather rigid and bristle-like, lower leaves much longer

Panicles ¾'-2' long, terminal and lateral, very narrow and contracted,
   lateral panicles enclosed in the sheaths   Spikelets 1-flowered,
   1½''-2'' long   Scales 3, outer scales slightly unequal, very acute, flowering scales very acute, rough toward the apex   Palet very acute, as
   long as flowering scale.   Stamens 2 or 3, anthers dark red, large
   Stigmas white

Dry and sandy soil.   August and September.
Vermont to Wyoming, south to Georgia and Texas

Gauze-grass.   *Sporóbolus uniflórus* (Muhl.) Scribn & Merr

Perennial

Stem 6'-18' tall, very slender, somewhat flattened, erect.   Ligule very
   short.   Leaves 2'-6' long, about ½'' wide, flat.

Panicle 3'-8' long, very delicate, branches hair-like, erect or ascending
   Spikelets 1-flowered, (occasionally 2-flowered), minute, about ⅝'' long,
   light purple, on slender pedicels.   Scales 3; outer scales obtuse, about
   equal, half the length of the acute flowering scale.   Stamens 2 or 3

Wet, sandy soil   July to September
Maine to Michigan, south to New Jersey.

THE BENT-GRASSES

BROWN BENT-GRASS, THIN-GRASS, ROUGH HAIR-GRASS, AND RED-TOP

"Soon will the high midsummer pomps come on"

Neither the earlier nor the later grasses so monopolize field and wayside as do those of this genus, whose hundred species are scattered through all the temperate regions of the world   As typical of midsummer warmth as goldenrod is of the largesse of autumn, they begin to bloom by the waysides of June, and symbolize, as do no other plants, the heat of summer with its hay fields and the endless, iterant call of the cicada.

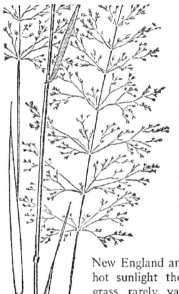

In drier places Brown Bent-grass (*Agróstis canìna*) is often found in bloom a month before the common Red-top, and it is also frequently seen in moister meadows as a red-brown mist closery following the blossoming of Velvet Grass. In bloom the plant calls to mind a miniature Red-top, but the leaves are narrower than those of the latter species, the basal leaves being almost bristle-form, while the flowering scale differs in developing a dorsal awn. Brown Bent is often seen on lawns and it is also quite common near the coastwise marshes of New England and New Jersey, where under the hot sunlight the widely open panicles of this grass rarely vary in colour from brown or brownish purple, flecked with white by the small anthers.

Thin-grass (*Agróstis perénnans*) is well described by its common name. The panicles are very pale green, rarely tinged with purple, and the short branches, with the branchlets and pedicels, are widely spreading. The whole plant is weak and slender, and the tiny flowers, opening soon after Brown Bent blossoms, are in outward appearance not unlike those of a small Red-top that has lost its colour through growing in a shaded place, but in examining a blossom with the microscope the palet is seen to be minute or lacking. Thin-grass is most frequently found in the damp soil of shaded pastures, and it is one of the comparatively few grasses that ascend the highest mountains of the Appalachians.

Red-top *Agrostis alba*

Among our common wayside grasses there are few more beautiful than the Rough Hair-grass, with its shining stems and wonderfully delicate panicles which glisten in the sunlight like purple cobwebs. When the Fescues are past, and the Redtop is in its glory of midsummer colouring, the slender stems of this grass droop by the wayside and may be passed a score of times unnoticed, for, although the flowering-heads are often a foot and a half long and half as broad, the widely spreading branches are so infinitely fine that the panicles seem to have gathered "fern-seed," since they so nearly "walk invisible." To see the plant in its greatest beauty one should seek an upland plain where the landscape gardening of Nature has planted the dark green of bush-clover and tick-trefoil against the summer grasses. Here, where the Wild Oat-grass was earlier abundant, and where later the Beard-grasses will endure throughout the autumn, are large tufts of Rough Hair-grass — the whole flowering-head, stem, branch, and spikelet, burned to rose-purple by the July sun. Before the panicles expand they are sometimes gathered and sold as "Silk-grass," but the name of Fly-away Grass is more appropriate as the seeds ripen, for the light panicles are soon broken by the wind and drift over the fields as the earliest tumble-weed.

Rough Hair-grass
Agrostis hyemalis

The Red-top and its varieties are among the

chief grasses of July fields, and in midsummer acre upon acre is clothed by them in varying tones of reddish purple. Perhaps native in the North and Northwest, Red-top was brought to the Eastern States from Europe in the eighteenth century and was cultivated as "English Grass," but its various uses and appearances under many conditions of soil and climate have given it a multitude of names. As Bonnet-grass it was common along the valley of the Connecticut River, where the stems were formerly cut to be braided into hats As Fioren a variety which produces a smooth and velvet-like turf was most highly extolled in England and Ireland as a winter fodder grass, and its sponsor was at one time caricatured as mowing grass in winter while snow lay upon the ground The most common form of Red-top is found in nearly all the states, and as in early summer the unopened panicles, in narrow spikes of green or purple, rise above the leaves the grass may be recognized several weeks before it is in bloom. Where the earth is moist the blossoms are darker in colour, and in the more luxuriant growth of rich soil rootstocks are formed, which in gardens are as difficult to eradicate as are those of the too-common Couch-grass

Red-top. Herd's Grass. Bent-grass. *Agróstis álba* L

Perennial   Exceedingly variable
Stem 1-4 ft tall, erect, not branched   Ligule 4″ long or less.   Leaves 2′-10′ long, 1″-4″ wide, flat, rough
Panicle 2′-10′ long, open, branches many   Spikelets 1-flowered, about 1″ long, green or reddish purple   Scales 3; outer scales acute, about equal, rough on keels, flowering scale obtuse or acute, palet not less than one third as long as flowering scale   Stamens 3, anthers white, short.
Fields, meadows, and waysides.   June to September
Throughout nearly the whole of North America

Rough Hair-grass. Fly-away Grass. *Agróstis hyemàlis* (Walt.) BSP.

Root biennial
Stem 1-2½ ft tall, erect, slender, not branched   Ligule 1″-2″ long   Leaves 2′-6′ long, ½″-1½″ wide, rough, basal leaves usually involute and bristle-form
Panicle 6′-20′ long, widely open, usually reddish purple, branches many, hair-like, rough, divided near the middle, spikelet bearing at the extremities, lower branches 3′-8′ long   Spikelets 1-flowered, 1″

long or less. Scales 3; outer
scales nearly equal, acute,
rough on keels; flowering
scale obtuse; palet minute.
Stamens 3, anthers small.
Dry or moist soil. June to
August.
Throughout nearly the whole
of North America, except
in the extreme north.

BLUE-JOINT GRASS AND
NUTTALL'S REED-GRASS

Though that "bank where
the wild thyme blows" be in-
accessible, the country holds many
a marshy meadow wherein all manner of delightful acquaintances may be
made. In such "marish places" grow
pitcher-plants, dotting the swale with
fairy parasols of rose and maroon,
orchids, fragile and beautiful in pink and
lavender, while treacherous sundews,
plants of doom to the lesser members of
the insect kingdom, are scattered among
the sedges and rushes above which rise
the taller grasses of moist grounds.

In June when the season is at its
height — though in reality it is only ap-
preciation that is more vivid in early
summer, for each week brings new bloom
and colour to the marsh — the Blue-joint
often covers large areas, or appears in
isolated specimens among the sedges.
This grass is tall and slender, bearing
narrow flowering-heads which are usually
strongly tinged with bluish purple, and
on some soils the dark green leaves
change to a dull purplish colour that is
noticeable even from a distance. The seeds ripen early, and
in some localities the grass is a difficult one to find in bloom, as

Blue-joint Grass
Calamagrostis canadensis

the flowers fade rapidly, leaving only pale-brown panicles of ripening seeds. Occasionally the whole panicle fails to mature, and then the spikelets remain empty and faded.

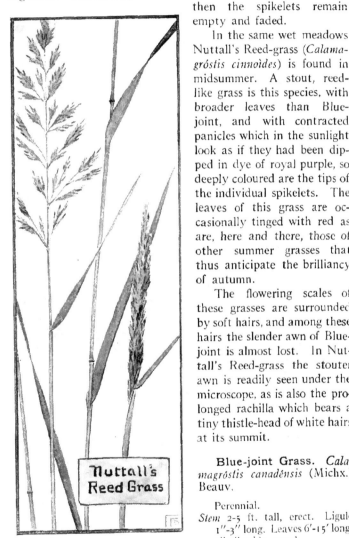

Calamagrostis cinnoides

In the same wet meadows Nuttall's Reed-grass (*Calamagróstis cinnoìdes*) is found in midsummer. A stout, reed-like grass is this species, with broader leaves than Blue-joint, and with contracted panicles which in the sunlight look as if they had been dipped in dye of royal purple, so deeply coloured are the tips of the individual spikelets. The leaves of this grass are occasionally tinged with red as are, here and there, those of other summer grasses that thus anticipate the brilliancy of autumn.

The flowering scales of these grasses are surrounded by soft hairs, and among these hairs the slender awn of Blue-joint is almost lost. In Nuttall's Reed-grass the stouter awn is readily seen under the microscope, as is also the prolonged rachilla which bears a tiny thistle-head of white hairs at its summit.

Blue-joint Grass. *Calamagróstis canadénsis* (Michx.) Beauv.

Perennial.
Stem 2-5 ft. tall, erect. Ligule 1″-3″ long. Leaves 6′-15′ long, 1″-4″ wide, rough.

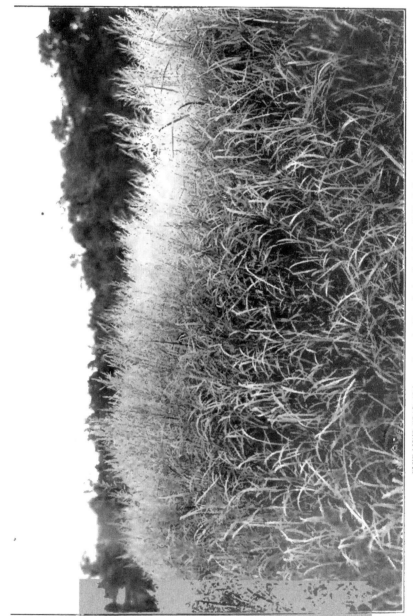

BLUE-JOINT GRASS (Calamagrostis canadensis) growing luxuriantly in Renwick Marsh, Cayuga Lake

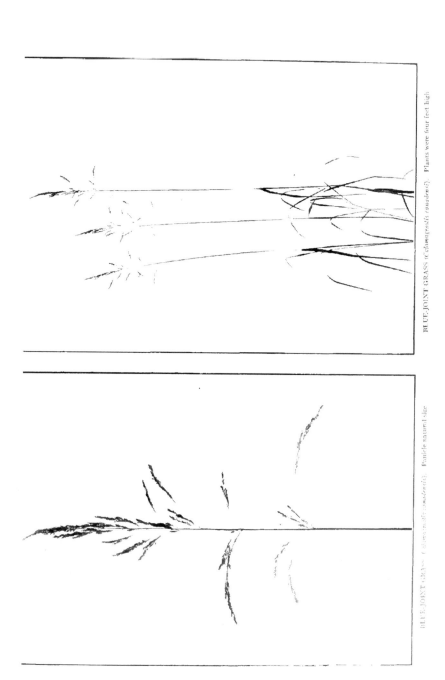

BLUE-JOINT GRASS (*Calamagrostis canadensis*). Panicle natural size.

BLUE-JOINT GRASS (*Calamagrostis canadensis*). Plants were four feet high.

Panicle 4'-10' long, oblong, often reddish purple, branches ascending or spreading. Spikelets 1-flowered, 1¼"-2" long. Scales 3: outer scales nearly equal, acute, rough; flowering scale thin and translucent, divided at apex, surrounded by numerous white, silky hairs which rise from the base of the scale, flowering scale bearing a dorsal awn about the length of the basal hairs. Rachilla prolonged and hairy. Stamens 3.

Wet grounds. June to August.

Newfoundland to Alaska, south to North Carolina and southern California.

MARRAM GRASS

"Where the gray beach glimmering
runs, as a belt of the dawn."

From Virginia northward, along the Atlantic coast, the gray-green leaves of Marram Grass add their subdued colour to the pale sands. Where the forests advance toward the water's edge this grass occupies a narrow strip of shore between tides and trees; but where the sands have drifted inland, driving vegetation before them, Marram Grass often covers large areas and aids in arresting the encroaching desolation. Although other plants are smothered by wind-blown sand, this grass, continuing to grow as the sand collects around it, is found on high dunes with whose rise it has kept pace. The buried stems attain an incredible length, and the stout rootstocks, becoming matted and interwoven, prevent the drifting of sands, and resist the action of waves and winds which wage unceasing warfare against the land.

The value of Marram Grass as a sand-binder has long been recognized. Even in the reign of Elizabeth laws were passed in England for the preservation of this grass, and in America, in colonial days, the in-

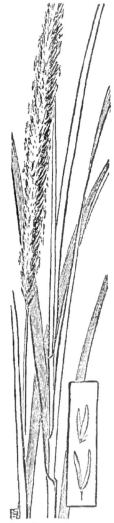

Marram Grass
Ammophila arenaria

habitants of certain towns in Massachusetts were obliged to plant Marram Grass each April, or suffer the penalty of their disobedience to law. It is said that the harbour of Provincetown, Mass, owes its preservation to the Marram Grass committee which was authorized to demand the cultivation of this grass along the coast

Great has been the devastation caused by sand-storms on coasts where there are no sand-binding grasses Following the thoughtless pulling up of Marram Grass on a shore of Scotland, such a storm in the winter of 1769 was so destructive that apple trees, it is said, were buried and only their highest branches left above the surface of the drifts

The long leaves of Marram Grass, or Beach Grass, as it is often called, are smooth on the outer surface, are finely ribbed within, and become involute in drying. The inflorescence is a cylindrical, spike-like panicle, composed of many one-flowered spikelets which in bloom are fringed with white anthers. The grass may easily be recognized, even from a distance, by the characteristic colour of its leaves, so perfectly does the silver green accord with the silver sands

Marram Grass. Beach Grass. Sea Sand-reed. *Ammóphila arenària (L)* Link.

Perennial, from extensively creeping rootstocks.

Stem 2-4 ft tall, stout, rigid, erect. Ligule a minute ring. Leaves 6'-24' long, 2"-6" wide, gray-green, smooth on lower surface, ribbed and rough on upper surface, soon involute.

Spike-like Panicle 5'-14' long, cylindrical, green, densely flowered, 5"-9" in diameter Spikelets 1-flowered, 5"-6" long Scales 3, compressed, outer scales about equal, acute, flowering scale nearly as long as empty scales and bearing a tuft of short hairs at the base, palet slightly shorter than flowering scale Rachilla prolonged Stamens 3, anthers white.

Sandy beaches along the coast July to October.

New Brunswick to Virginia, also on the shores of the Great Lakes, and in California.

WOOD REED-GRASS AND SLENDER WOOD REED-GRASS

Leafy stems of Wood Reed-grass arrest the attention before the ample panicles are visible, for, although this grass does not

MARRAM GRASS (*Ammophila arenaria*) growing in drifting sand. Seaside goldenrod (*Solidago sempervirens*) is also striving to keep its leaves above the drifts

bloom until late summer, the reed-like stems, frequently shoulder-high, and bearing broad, soft leaves, are common in July in wooded swamps and by shaded streams.

When the many-flowered panicles first appear they are pale green, contracted, and almost silky; later, as the flowers open, the multitude of hair-like branches spread from the flowering-head and the spikelets are often tinged with purple, while as the seeds ripen the panicles are again contracted as before blossoming.

Slender Wood Reed-grass (*Cinna latifòlia*) is also a grass of late summer, and though in many localities this species is less common than Wood Reed-grass it is not infrequently found in deep woods and on mountainsides. It is more slender than the preceding species and bears a less densely flowered panicle of spreading or drooping branches.

The flowers of these grasses have but one stamen, and the palets are remarkable in that they show but one nerve.

Wood Reed-grass. *Cinna arundinàcea* L.

Perennial.

Stem 2-6 ft. tall, leafy, not branched, erect. Ligule 1"-2" long. Leaves 6'-15' long, 3"-7" wide, flat, roughish.

Panicle 6'-15' long, densely flowered, rather narrow, green or purple. Spikelets 1-flowered, about 2½" long. Scales 3; outer scales rough, unequal, acute; flowering scale 2-toothed, usually bearing a minute awn between the teeth. Rachilla sometimes slightly prolonged. Palet 1-nerved. Stamen 1.

Moist woods, thickets, and swamps. July to September.

Newfoundland to the Northwest Territory, south to Alabama and Texas.

Wood Reed-grass
Cinna arundinacea

131

SILVERY HAIR-GRASS

Silvery Hair-grass is a tiny annual that has become only locally abundant since its accidental introduction from Europe. It is occasionally found in the sandy soil of waste fields, and in dry places by the wayside, where the Slender Fescue strives with small success to draw life from the unpromising ground. Silvery Hair-grass is of slender growth, and its bristle-like leaves resemble those of a small plant of Wavy Hair-grass, which blooms a few weeks later.

Sand-growing annuals, like those near the deserts, are of rapid growth, and take advantage of a spring shower to grow, bloom, and mature their seeds in as short a time as possible. Therefore the Silvery Hair-grass early lifts its spreading panicles from the ground and opens the purplish spikelets for a day. It is rarely a foot in height, and often much less than that, one of the smallest of the grasses that bloom in the Eastern States, and silvery, as its name implies, as the colour fades from the blossoms, and the empty scales, shining and translucent, remain on the panicles long after the ripened seeds, with their adherent flowering scales, have floated away on the breeze.

Silvery Hair-grass. *Aira caryophyllèa* L.

Annual. Naturalized from Europe.
Stem 4'-12' tall, slender, erect. Ligule about 1" long. Leaves bristle-form, ½'-2' long.
Panicle 1'-4' long, very open, branches hair-like, spikelet-bearing toward the extremities. Spikelets 2-flowered, 1"-1½" long, green and rose-purple, turning silvery and translucent in fading. Scales 4; outer scales acute, equal; flowering scales acute, 2-toothed, awned; awns slender, 2" long or less. Stamens 3.
Dry soil and waste places. May to July.
Massachusetts to Virginia, also on the Pacific Coast.

Silvery Hair-grass
Aira caryophyllèa

VELVET GRASS

"I find myself 'mid pastures sweet,
Vernal, green, and ever gay."

The eye is arrested in mid-June by bits of colouring in the fresh meadows, as if on the darker grasses a grayish-pink fog rested, clinging in unevaporated clouds where the Velvet Grass blooms. By waysides and in meadows the soft panicles open, white and gray-green, pale pink and purple, charming in colour, and surely more beautiful than the Yorkshire fog from which the English named the grass.

Soft white hairs clothe leaves and sheaths in a dense pubescence, and from this alone the grass may be recognized throughout the season, for although the ripening stems change to shining yellow the sheaths remain green and retain their velvet-like softness until fall. The plant is usually about two feet high, though in the Southern States it is often much taller. The long upper sheath is inflated and until the flowers open it encloses the soft panicle which, though richly coloured where the sunlight touches it, is sometimes pale greenish white, even in bloom. Against the green of early summer grasses the flowers of Velvet Grass are very noticeable, and until September a few plants still bloom and may be found in many locations from the borders of damp thickets to sandy fields and shores.

Velvet Grass
Holcus lanatus

Doctor Muhlenberg, who did much to bring before the world the agricultural resources of our country, termed this grass "excellens pabulum," but cattle are not fond of Velvet Grass and farmers do not find it worthy of cultivation, save on poorest soils where more desirable species fail to thrive.

Velvet Grass. Salem Grass. Yorkshire Fog. *Hólcus lanátus* L.

Perennial, with creeping rootstocks   Naturalized from Europe.

Stem 1-3 ft. tall, erect   Sheaths and leaves clothed in soft white hairs. Ligule about 1″ long or less   Leaves 1′-7′ long, 2″-6″ wide, flat, very soft, grayish green

Panicle 1′-6′ long, pyramidal, open in flower, downy, greenish white, tinged with pink, rose, or purple. Spikelets 2-flowered, about 2″ long, lower flower perfect, upper flower staminate. Scales 4; outer scales slightly unequal, clothed in short white hairs, 1st scale acute or obtuse, 1-nerved; 2nd scale awn-pointed, 3-nerved, flowering scales papery, smooth, the 1st obtuse, the 2nd 2-toothed and bearing from just below the apex a short awn which soon becomes hooked. Stamens 3, anthers white, yellow, or lavender.

Meadows, waysides, and waste places.   May to August.

Nova Scotia to Ontario and Illinois, south to North Carolina and Tennessee.

MARSH OATS, MEADOW SPHENOPHOLIS, SLENDER SPHENOPHOLIS, AND EARLY BUNCH-GRASS

In the crevices and depressions of those rocks that push out into woodland brooks, mosses and lichens surround the roots of a few plants which might appear to be true air-plants, deriving their sustenance from the winds of heaven, so scanty are the visible means of support. But pull up the stalk of one of these—flower, and stem, and root—and see how closely the rootlets hug the rock, and penetrating the tiniest crevices hold the plants as securely as though they were anchored on deep soil.

One of the early summer grasses on such rocks, and near them, is the Marsh Oats, which, although allied to other Oat-grasses, unlike them is found in the low grounds of wet meadows and by brooksides. The grass is slender, with thin, flat leaves, and narrow, loosely flowered panicles whose flat spikelets of pale green and yellow bear each a conspicuous awn from the upper flower. In this latter peculiarity it differs from other species of this genus which, like it, are slender, light green grasses of early summer.

Although natives of this country these grasses are seldom found in great abundance in the East   Meadow Sphenopholis, a tufted perennial, with narrow panicles of pale green spikelets, is sometimes common in moist woods and meadows, but it is a plant little noticed save by the student to whom each new grass is a discovery of absorbing interest.

Dry, open woods are the home of the Slender Sphenopholis (*Sphenópholis nítida*) which shows a more open panicle in bloom, though the branches are drawn closely to the stem before and after blossoming. The plant is somewhat more slender than others of this genus, and the spikelets are not so crowded, but the same pale green colour is characteristic of the whole plant.

Early Bunch-grass (*Sphenópholis obtusàta*) blooms in dry soil, and when in bright sunlight the spike-like panicles of short, erect, densely flowered branches are frequently tinged with greenish purple. This grass soon fades, and in July the slender stems become a shining yellow tinged with pink.

Marsh Oats. *Sphenópholis palústris* (Michx.) Scribn.

Perennial.
Stem 1-3 ft. tall, slender, erect. Ligule less than 1″ long. Leaves 1′-5′ long, about 2″ wide, flat, roughish.
Panicle 2′-8′ long, narrow, loosely flowered, yellowish green, branches short, slender. Spikelets 2-flowered, about 3″ long. Outer scales acute, nearly

Marsh Oats
Sphenopholis palustris

Meadow Sphenopholis
Sphenopholis pallens

135

equal; flowering scales roughish, 2-toothed, scale of lower flower awnless (or rarely bearing a rudimentary awn), scale of upper flower bearing from between its teeth an awn about 3" long   Rachilla prolonged   Stamens 3

Low grounds   May to July.

Massachusetts to Illinois, south to Florida and Louisiana.

Meadow Sphenopholis.   *Sphenópholis pállens* (Spreng.) Scribn.

Perennial, tufted

Stem 1-3 ft tall, slender, erect   Ligule 1" long or less.   Leaves 2'-7' long, 1"-3" wide, rough, flat, pale green

Panicle 3'-8' long, narrow, pale green.   Spikelets 2-3-flowered, nearly 2" long   Outer scales unequal, 1st scale acute, very narrow, about ¼ the width of the 2nd scale, which is obtuse or abruptly acute, flowering scales acute   Stamens 3.

Damp woods and meadows   May to July

Maine to Wisconsin, south to North Carolina and Texas

WAVY HAIR-GRASS AND TUFTED HAIR-GRASS

By the dry paths of early summer delicate panicles of Wavy Hair-grass rise in silvery pink and green above thread-like leaves. The flowering-heads of this grass are widely open, and as the small spikelets are borne only at the extremities of the wavy branches the plant seems but a transient spirit of the wayside that must be begged to tarry lest it leave ere

> "the hasting day
> Has run
> But to the even-song"

The silvery scales are exquisitely tinted in pink and rose for the short time that the flowers are open, but as the flowers fade the scales lose their colour and persist, shining and translucent, long after the seeds have ripened and the stems have died   This grass grows in the shade of wooded pastures as well as in the sunlight, where on dry hillsides dark green tufts of the involute root-leaves are frequently seen   Wavy Hair-grass is found at higher altitudes than are many of the common grasses, and in spring it is the most slender species in blossom until the misty panicles of Fly-away Grass open, wraith-like in their beauty.

Tufted Hair-grass (*Deschámpsia caespitósa*) prefers the moister soil of lake shores and river banks, where the tall stems bear widely

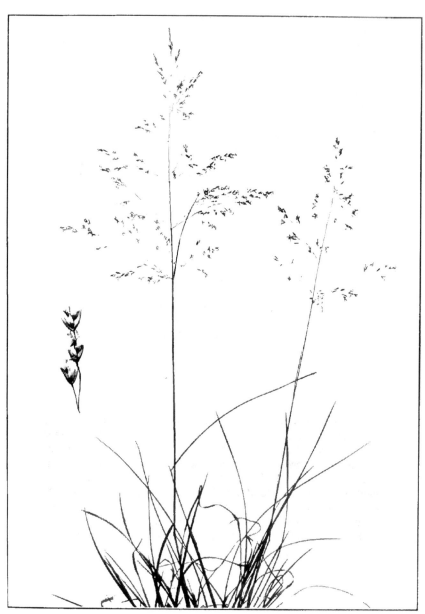

WAVY HAIR-GRASS (*Deschampsia flexuosa*) One-half natural size Spikelets enlarged by two

open panicles tinged with blue and purple. This grass is spoken of by one writer as among the tallest of British grasses, often attaining a height of six feet in that ᴉntry, where its stems are occasionally used in weaving coarse floor-mats. Here, the Tufted Hair-grass is from two to four feet tall, a variable species with flat leaves, somewhat smaller spikelets, and stouter stems than are seen in Wavy Hair-grass.

Wavy Hair-grass. *Deschámpsia flexuòsa* (L.) Trin.

Perennial, tufted.

Stem 1-2½ ft. tall, slender, erect. Sheaths much shorter than internodes. Ligule 1″-2″ long. Leaves 1′-7′ long, involute and bristle-like, those of the stem very short.

Panicle 2′-8′ long, widely open, branches hair-like, spreading, wavy, spikelet-bearing toward the extremities. Spikelets 2-flowered, about 2½″ long, green tinged with rose, silvery and translucent in fading. Scales 4; outer scales acute, slightly unequal; flowering scales acute, divided at apex, hairy at base, bearing a bent and twisted dorsal awn about 3″ long. Rachilla prolonged. Stamens 3.

Dry soil. June to August.

Labrador to Ontario, south to North Carolina and Tennessee.

OATS, CULTIVATED AND WILD

"It is their chiefest bread-corn for Iannocks, Haver cakes, Tharffe cakes, and those which are called generally Oten cakes: and for the most part they call the graine Haver, whereof they do likewise make drinke for want of Barley."

Who does not remember a field of oats under a July sky? The bright leaves

"Green, and all of a height, and unflecked with a light or a shade."

Wavy Hair-grass
Deschampsia flexuosa

One of the grains cultivated in ancient times, this is still the principal cereal of the extreme north of Europe, where the grain is used daily in the kitchens of the working people, who from it make their "fladbrode," or oaten cakes.

Many varieties of this cultivated grass (*Avèna sativa*) have been developed in more modern days, the chief forms of which are Panicled Oats and Banner Oats — the former with symmetrical, and the latter with one-sided panicles — and the large, drooping flowering-heads frequently bloom by the waysides where seed has been accidentally dropped.

Although there are many species in this genus, none of them is common in the Eastern States. The Wild Oats (*Avèna fátua*) is, perhaps, the most interesting, but it is not often found east of the Mississippi. The peculiar flowering scales of this grass are half an inch long and are covered with stiff, brown hairs from among which projects a bent and twisted awn more than an inch in length. Like the twisted awns of other grasses, the awn of Wild Oats, when dampened, quickly uncoils and moves in a most weird manner, as if suddenly endowed with life.

It may be that the apparent inward volition of this strange awn was looked upon as an evidence of influence from the under-world, for it is certain that the term "wild-oats" has long been a synonym for the worthless — and enjoyable. Where the Norse mythology credited the dwarf

Cultivated Oats

140

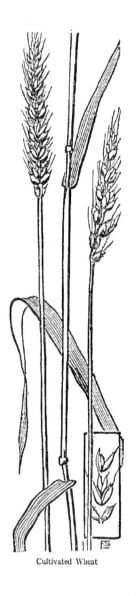

Cultivated Wheat

Cultivated Rye

141

Loki with troubling poor mortals by his evil deeds, that grass was called "Loki's Grass," or "Dwarf's Grass," the common proverb

Meadow Oat-grass
Arrhenatherum elatius

about the scapegrace being: "Loki is sowing his seed in him." Jugglers of former days used the awn of Wild Oats in foretelling future events, and imposed on the credulous by calling it "the leg of an enchanted fly" or "the leg of an Arabian spider." Surely such pseudo-sciences as palmistry and astrology would lose their charm when one could place unwavering faith in the fortune-telling based on the strange movements of this enchanted awn.

An English book speaks of the use of the flowering scale as an artificial fly in trout fishing. The twisted awn, uncoiling as it is dropped in water, whirls the hairy scale about and causes it to appear like a struggling insect.

MEADOW OAT-GRASS

For some long lost reason the unusual name of "Grass of the Andes" was once given to the Meadow Oat-grass which in early spring is often found growing in loose tufts near fields and hedges. A rapid and rigidly erect growth soon lifts the narrow blossoming-heads of this plant above the leaves of later flowering Fescues and Bent-grasses, and in bloom a rare combination of colouring is shown in the brownish-green spikelets and yellow anthers.

As an important meadow grass this species was introduced from Europe many years ago, yet when one of the earlier American writers on agriculture speaks of Meadow Oat-grass as being

142

cultivated "by a few curious farmers" he leaves the interpretation of the adjective to the prejudices of his readers.

The long, fibrous roots, on which bulbous formations are occasionally developed, give to the plant great drouth-resistant qualities, but, though valued in the South and extreme West, Meadow Oat-grass has hardly proved itself worthy of extensive cultivation north of Mason and Dixon's line.

Meadow Oat-grass. *Arrhenathèrum elàtius* (L.) Beauv.

Perennial.

Stem 2-4 ft. tall, erect. Ligule about 1" long. Leaves 3'-12' long, 1"-4"wide, flat, rough.

Panicle 4'-10' long, narrow, branches short, erect or ascending. Spikelets 2-flowered, 3"-5" long, brownish, lower flower staminate, upper flower perfect. Scales 4; outer scales acute, unequal; flowering scales sparingly hairy, scale of lower flower bearing a bent and twisted dorsal awn about 6" long, scale of upper flower bearing a very short, straight awn between its teeth. Rachilla prolonged. Stamens 3, anthers yellow. The fresh plant has a decidedly bitter taste.

Fields, waysides, and waste places. May to August.

Maine and Ontario to Georgia and Tennessee, also on the Pacific coast.

WILD OAT-GRASS, FLATTENED OAT-GRASS, AND SILKY OAT-GRASS

In rock-strewn pastures, where the scanty soil supports low sumac and fragrant bayberry, a slender, wiry grass covers the dry knolls and blossoms in early spring soon after Sweet Vernal blooms in moister fields. This grass, so common on poor soil from Canada to the Gulf States, is Wild Oat-grass, a species that varies not only in size but also by occasionally clothing with silky hairs its lower leaves. The larger plants differ little from a small growth of Flattened Oat-grass (*Danthònia compréssa*) which blooms a few weeks

Wild Oat-grass
Danthonia spicata

143

later and has a more limited range Both of these grasses have narrow leaves and are easily recognized, as the short, few-flowered panicles bear but few branches and are unlike those of other grasses of early summer In Flattened Oat-grass, which is especially abundant in mountain meadows and ascends the highest peaks of the Appalachians, the stems are flattened and the flowering scales terminate in longer and more slender points than do those of Wild Oat-grass

Silky Oat-grass (*Danthònia sericea*), another early species of sandy soils, is less common in the Northern States than are the other two grasses. It is slightly stouter than either and is usually very silky on leaves and sheaths, while the flowering scales are whitened with soft hairs.

Wild Oat-grass. *Danthònia spicàta* (L) Beauv.

Perennial

Stem 1-2½ ft tall, slender, erect. Lower sheaths often downy. Ligule very short. Leaves 4'-6' long, 1" wide or less, often involute.

Panicle 1'-2¼' long, branches few. Spikelets 5-8-flowered, 4"-5" long, green, few Outer scales long and narrow, smooth, usually extending beyond the uppermost flower, flowering scales broad, 2-toothed, downy, bearing from between the teeth a bent and twisted, spreading awn about 4" long. Awn purple in the twist, green above. Stamens 3.

Dry and rocky soil. May to August

Newfoundland to Dakota, south to Florida and Texas.

THE SPARTINAS

FOX-GRASS, CREEK SEDGE, CORD-GRASS, AND SALT REED-GRASS

"By a world of marsh that borders a world of sea."

Spartinas are lovers of strong salt breezes, and with Marram Grass and Bitter Panic-grass are characteristic plants of sea sands and salt marshes. With long rootstocks, sharp-tipped with scaly points, the grasses of this genus grow to the water's edge, and withstand the daily flooding of the tides. Pull up a portion of the strong rootstock from some large Spartina and the growing point is sharp as any needle. These subterranean stems, pushing their way in endless, interlacing network through the sands and creek mud, aid in firmly binding the unstable shores.

Fox-grass, the earliest, and by far the most slender of the common species, blooms through July and August and covers large

CREEK SEDGE ON THE MARSH (*Spartina glabra*, var. *pilosa*)

CREEK SEDGE IN BLOOM (*Spartina glabra*, var. *pilosa*)

BLOSSOMING SPIKES OF CREEK SEDGE (*Carex* ... ... on the left the pistils are mature, on the right the stamens

SALT REED-GRASS (*Spartina cynosuroides*) by the ditch in the marsh

SALT REED-GRASS

FOX-GRASS (*Spartina patens*). Soil and fresh water coming down from the bank, have made this rank growth

"A LEAGUE AND A LEAGUE OF MARSH-GRASS." Fox-grass, as far as the eye can see. Locally called "White Salt." (*Spartina patens*)

FOX-GRASS

E MARSH HAS BEEN COVERED WITH SAND AND THROUGH IT STRUGGLE STRAY PLANTS OF FOX-GRASS.
(Spartina patens)

FOX-GRASS. *Vilfa vaginæflora.* Natural size.

areas with its smooth, dark green leaves and stems and its vividly coloured flowering spikes of purple flecked with yellow anthers. This and the Black-grass — really a rush — are valuable plants of the seaside, and their low, dense growth of leaves and stems, so frequently seen by tidewaters, is easily recognized by the characteristic dark green colour. These plants yield a large amount of the salt hay gathered each year, and the wiry stems of Fox-grass are much used in packing. This, also, is one of the many plants that have been proved to yield fibres suitable for spinning and weaving, and it is said that although the fibre from the stems of Fox-grass is deficient in length it is equal in strength and fineness to that of flax.

Creek Sedge advances into the water and one must wait until low tide before approaching by land the pale green spikes which are so beautifully fringed with their white anthers. This grass is common everywhere along our coasts and by creek margins, where it borders the water with impassable thickets of reed-like stems which remain green until late autumn. The stout stems, leafy to

Fox-grass
Spartina patens

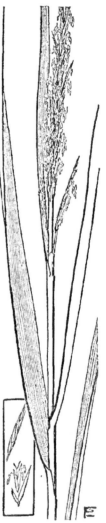

Creek Sedge
Spartina glabra, var. pilosa

their summits, are sometimes used for thatching, but the plant, until thoroughly dry, has an unpleasant odour

Cord-grass (*Spartina Michauxiana*), found by river and lake borders, as well as near the coast, is a tall, stout grass whose smooth stems bear an inflorescence composed of five to thirty light-coloured, erect or spreading spikes Paper and twine have been manufactured from this species, and, like Creek Sedge, Cord-grass is used to form a waterproof thatch.

Salt Reed-grass (*Spartina cynosuroides*) is the largest of the genus and bears broad, rough leaves, and a dense inflorescence of many spikes, which are often tinged with purple On the Jersey marshes, acres are covered by this grass, which sometimes attains a height of ten feet, with stems an inch in diameter at the base.

Fox-Grass. Salt-meadow Grass. *Spartina pàtens* (Ait) Muhl.

Perennial from creeping rootstocks.

Stem 1-3 ft. tall, slender, erect, sometimes reddish.   Sheaths and lower surface of leaves very smooth. Ligule a ring of very short hairs.   Leaves 4'-14' long, 1''-2'' wide, dark green, involute, rather rigid

Spikes 2-7, alternate, narrow, 1-sided, 1'-2' long, erect or spreading, on short pedicels. Spikelets 1-flowered, narrow, 4''-5'' long, in 2 rows, green or pinkish purple Scales 3, outer scales acute, unequal, rough on keels, flowering scale slightly 2-toothed, palet slightly longer than flowering scale Stamens 3, anthers purple or reddish brown

Salt meadows and sandy shores along the coast.   June to September. Newfoundland to Virginia.

Creek Sedge. Creek Thatch. Smooth Marsh-grass. *Spartina glàbra* Muhl. var. *pilòsa* Merr.

Perennial, from stout, creeping rootstocks.

Stem 2-9 ft tall, stout, reed-like, erect   Sheaths and leaves smooth. Ligule a ring of short hairs   Leaves 5'-24' long, 2''-8'' wide

Spikes usually many, 1'-4' long, erect, 1-sided, forming a terminal, spike-like inflorescence Spikelets 1-flowered, narrow, 6''-8'' long, in 2 rows   Scales 3; outer scales acute, unequal, rough on keels, palet slightly longer than flowering scale   Stamens 3, anthers light coloured   Stigmas long, white A variable species

In salt marshes and by the borders of creeks.   July to October On the Atlantic and Pacific coasts.

BERMUDA GRASS

In the beauty of old mythology, when the white clouds were cattle driven through the wide pasture of the sky, the sacred Vedas celebrated Bermuda Grass as the "Shield of India," and "Preserver of Nations," a plant sacred to the Hindoos, as without it the herds would perish and famine consume the people. The twentieth century, with face turned traffic-ward, notices Bermuda Grass merely as "the most valuable forage grass of the Southern States."

A lover of heat and sunshine, this grass is seldom found in the North, but from southern New York State to the Gulf the short, finger-like spikes are not uncommon in dry soil where the low stems rise bearing narrow leaves which are crowded at the base of the stems and on the prostrate runners. The blossoming spikes are painted with dark purple stigmas, and Sir William Jones, in his "Asiatic Researches," long ago praised the extraordinary beauty of the flowers. Spreading extensively over the surface of the ground, the plant is tenacious of life, even during the driest seasons, and is highly valued as a lawn-grass in the South. Seed is rarely produced north of the Gulf States, but the runners often grow over rocks six feet across, or down precipitous embankments, and are most useful in holding arid and shifting sands.

Bermuda Grass. Scutch-grass. *Cýnodon Dáctylon* (L.) Pers.

Perennial, extensively creeping. Naturalized from Europe.

Stem 6'-24' tall, erect. Ligule composed of soft hairs. Leaves 1'-2½' long, ½"-2" wide.

Spikes 3-5, narrow, 1'-3' long, spreading from the summit of stem. Spikelets 1-flowered, about 1" long, borne on one side of the spike.

Bermuda Grass
Cynodon Dáctylon

Scales 3; outer scales acute, slightly unequal, rough on keels; flowering scale longer and broader; palet slightly shorter than flowering scale. Rachilla prolonged. Stamens 3. Stigmas purple.

Fields and sandy soil. June to September.
Southern New York to Florida and Texas.

TALL GRAMA

The Gramas, or Mesquites, are characteristic grasses of the Southwest, where they are a valued herbage of the ranges, but with the exception of two species that appear to have been introduced into Florida the Tall Grama is the only eastern member of the genus.

Tall Grama can hardly be classed among our common grasses of the Eastern States, yet the dense, leafy tufts are occasionally seen on dry hillsides and plains. The plant blooms in mid-summer and is easily recognized by the spreading or downward-pointing spikes of the long and narrow inflorescence, which for a short time is hung with anthers nearly as brilliant in colour as are the petals of the cardinal-flower.

Tall Grama. Side-oats. Racemed Bouteloùa. *Bouteloùa curtipéndula* (Michx.) Torr.

Perennial, tufted.

Stem 1-3 ft. tall, erect. Sheaths loose, sparingly downy. Ligule a ring of short hairs. Leaves 3'-12' long, 1''-2'' wide.

Spike 8'-15' long, somewhat 1-sided, composed of 20-60 spreading or downward-pointing spikes 3''-8'' long. Spikelets 1-flowered, 3½''-5'' long, in 2 rows on one side of the rachis. 4-12 spikelets in each spike. Outer scales roughish, acute, slightly unequal; flowering scales terminating in 3 short, awn-pointed teeth. Rachilla prolonged and bearing an awned rudiment of a second flower. Stamens 3, anthers red.

Dry soil. June to September. Ontario and Manitoba, south to New Jersey, Kentucky, Texas, and California.

Tall Grama
Bouteloua curtipendula

WIRE-GRASS AND EGYPTIAN GRASS

The coarse leaves and stems of Wire-grass form a thick, green carpeting in dooryards and by footpaths in many of the states during the middle and latter part of summer. This grass is a native of warm countries of the Eastern Hemisphere, and, although gradually becoming common in southern New England, is most abundant in the South, where it usually suffers the reproach of being called a weed, so rarely are plant immigrants honoured unless they pay the toll of usefulness.

Wire-grass is low and leafy, sending up numerous flowering-heads which in appearance call to mind the familiar Crabgrass, though the most casual observer could hardly mistake it for that species, since the spreading spikes of Wire-grass are so much heavier and thicker. On the "coasts of Coromandel" a stout species of this genus was cultivated for its large, farinaceous seeds which were used as food.

Egyptian Grass (*Dactyloctènium aegýptium*), whose grain has been used for food and also medicinally, resembles Wire-grass and is found in similar locations, though it is less widely distributed in this country. It blooms in late summer and from the preceding species it may be distinguished by the sharp terminal points of

Wire-grass
Eleusine indica

the spikes where the prolonged rachis extends beyond the uppermost spikelets.

Wire-grass. Goose-grass. Yard-grass. *Eleusîne indica* (L.) Gaertn.

Annual, tufted. Naturalized from Asia or Africa.

Stem 6′-24′ tall, flattened, erect or spreading. Sheaths loose. Ligule very short. Leaves 3′-10′ long, 1″-3″ wide, flat, rather thick.

Spikes 2-8, 1′-3′ long, spreading from the summit of the stem. Spikelets 3-6-flowered, 1½″-2″ long, in 2 rows on one side of the rachis. Outer scales acute, about equal; flowering scales acute. Stamens 3.

Cultivated grounds and waste places. June to September.

Southern New England to Ohio and Kansas, south to Florida and Texas.

SALT-MEADOW LEPTOCHLOA

The salt marshes and beaches — what wonderfully successful plants they contain, securely anchored, though on drifting sands, and braving the power of waves and winds! The greater number of our seaside plants bloom in late summer and spend their earlier strength in developing strong roots, which hold them firmly in place, and thick leaves which are unwithering beneath burning skies.

Like other flowering plants of the shore, all of the true salt-water grasses are late in blooming and bear coarse leaves that endure the lashing of storms. Salt-meadow Leptochloa is a low grass that grows in tufts in brackish marshes or meadows and also on saline soil toward the interior of the country. The stems are spreading, abundantly branched, and frequently send out roots from the basal joints. The leaves are long and narrow and the uppermost leaf encloses the base of the long panicle which is composed of erect, nearly sessile spikelets.

Salt-meadow Leptochloa
Leptochloa fascicularis

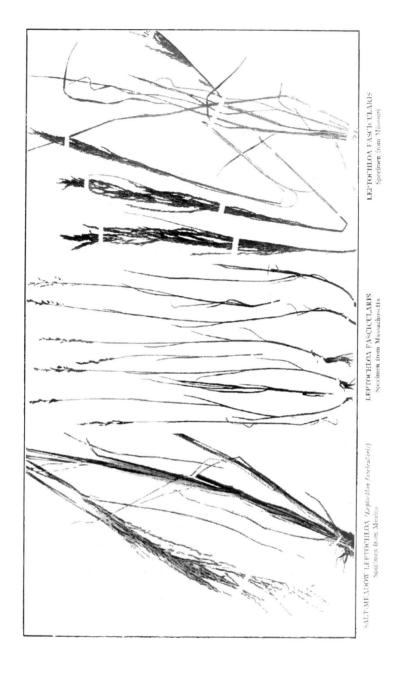

SALT-MEADOW LEPTOCHLOA (*Leptochloa fascicularis*)
Specimen from Mexico

LEPTOCHLOA FASCICULARIS
Specimen from Massachusetts

LEPTOCHLOA FASCICULARIS
Specimen from Missouri

**Salt meadow Lep-
tochloa.** *L\;Leptóchloa
fasciculàris* (Lam.)
Gray.

Annual, tufted.
Stem 1-2½ ft. tall, erect
or spreading, usually
much branched.
Sheaths loose, upper
one enclosing the base
of the panicle. Ligule
1″-2″ long. Leaves
3′-12′ long, 1″-3″
wide, flat.
Panicle 4′-12′ long,
composed of nu-
merous slender spikes on which
are borne the nearly sessile,
erect spikelets. Spikelets 5-
10-flowered, 3″-5″ long. Outer
scales unequal, acute, rough
on keels; flowering scales hairy
on margins near base, 2-tooth-
ed at apex and bearing a short
awn between the teeth. Stamens 3.
Salt marshes. July to September.
Southern New England to Florida and Texas,
also in saline soil from western New York to
Nevada and Mexico.

Reed
Phragmites communis

REED

This is one of the largest of our native
grasses. On the borders of ponds and in
marshes it forms tropic-like jungles of stout,
leafy stems that at last bear panicles of violet
and purple which change to plumes of silvery
white as the blossoms fade.

To brackish marshes along the coast the
Reed adds a wonderful beauty as in early
autumn the warm light of sunset steals over

"A league and a league of marsh-grass."

Tones of deep rose, lavender, and brown, that were unthought of
in the light of noontime, are brought out, and panicles of Reed,

rising above the tall marsh grasses, seem touched with frosted silver

This grass blooms in late summer but is not in its greatest beauty until September and October, as the silky hairs clothing the flowers are scarcely perceptible until the spikelets begin to ripen, then the hairs spread widely, and the leafy stems, sometimes more than ten feet tall, are surmounted by feathery panicles which resemble the soft plumes of Pampas Grass and remain until after the first snows.

The Reed, with its long rootstocks, is one of the many plants used by Nature as she slowly changes the land's surface, and transforms swamps and stagnant pools into fertile meadows. The horizontal rootstocks, spreading far from the stem, form a densely interwoven mat that holds mud and decaying vegetation and so affords a resting place for water-loving plants, which in turn are held and give firmer soil to marsh plants and grasses Day after day these transformations are in process around us, but so slowly does Nature perform her work that decades pass without appreciable change

The Old World has demanded more utilitarian service from her plants than we have, and although the Reed is sometimes cultivated in American gardens it is seldom used, as it is in Europe, to cover the roofs of farmhouses and outbuildings with a durable, water-proof thatch.

Reed. *Phragmites communis* Trin.

Perennial, from stout rootstocks.
Stem 5-15 ft. tall, stout, leafy, erect ˙ Sheaths loose. Ligule a ring of short hairs   Leaves 6'-24' long, $\frac{1}{2}$'-2' wide.
Panicle 6'-15' long, pyramidal, many-flowered. Spikelets 3-6-flowered, 5"-8" long, lowest flower often staminate Outer scales acute, unequal, flowering scales awl-shaped, pointed, thrice the length of the palet Rachilla bearing long, silky hairs which equal the flowering scales in length. Stamens 3
Borders of ponds and rivers, and in coast marshes. July to September
Throughout the United States and in southern Canada.

TALL RED-TOP

In dry fields, where the chief grasses of late summer are low, bushy Panic-grasses, slender Paspalums, and spreading tufts of Eragrostis, Tall Red-top, often shoulder-high and bearing long, tapering leaves, rises in striking contrast to the lower growth. This grass, which is found from southern New England to the Gulf, blooms in August and September, with the Purple Eragrostis, at a season when the sunshine brings from the earth the warm odour of pennyroyal and other mints that are common on dry hillsides, and that seem to have absorbed the summer's heat to give it out again in fragrance. The flowering-head of Tall Red-top, which is somewhat sticky to the touch in the axis of the panicle and below, is often more than a foot long and nearly as wide, and as the slender, rather rigid branches spread widely the panicles are very beautiful when the shining purple spikelets open.

Tall Red-Top. *Trìdens flàvus* (L.) Hitchc.

Perennial.

Stem 3-6 ft. tall, erect. Sheaths hairy at the summit. Leaves long, tapering, flat or sometimes involute.

Panicle 8′-20′ long, branches spreading, lower branches 3′-8′ long. Panicle sticky in axis and below. Spikelets purple, 4-8-flowered, 3″-4″ long. Outer scales unequal, keeled, abruptly pointed; flowering scales 3-nerved, slightly 3-toothed, nerves silky below. Stamens 3.

Dry fields. July to September.

Southern New England to Missouri, and southward.

SAND-GRASS

Sand-grass, thick and rigid of leaf, is a tufted plant of the beaches, and, like a few other salt-water grasses of the Atlantic coast, it is also found on Western ranges. It is very different in appearance from such species as Marram Grass, Bitter Panic-grass, and Creek Sedge, which also

Sand-grass
Triplasis purpurea

171

grow by the shore, but like them it does not bloom until late summer. Many stems, usually less than two feet tall, spring from one root and bear very short, narrow, rough leaves. The panicles, also, are short, with but a few stiff branches, and bear loosely flowered, rose-purple spikelets The outer scales are smooth but the flowering scales are fringed and bearded, presenting a distinguishing feature by which this grass may easily be recognized, and the acid taste of the plant is also peculiar to it

Sand-grass *Triplasis purpùrea* (Walt) Chapm.

Perennial, tufted

Stem 1-3 ft tall, erect or spreading Nodes usually downy. Ligule a ring of short hairs Leaves rigid, awl-shaped, $\frac{1}{2}'$-3' long, 1'' wide or less

Panicle 1'-3' long, branches few, at length spreading Lateral panicles usually included in the sheaths Spikelets 2-5-flowered, 2''-4'' long, loosely flowered Outer scales about equal, smooth, flowering scales very hairy on nerves, 2-lobed at apex and bearing a short, straight awn between the lobes, palets hairy on upper part of keels Stamens 3.

Sandy soil, especially along the coast July to September

Maine to Florida, westward to Nebraska and Texas

PURPLE ERAGROSTIS, LACE-GRASS, TUFTED ERAGROSTIS, PURSH'S ERAGROSTIS, STRONG-SCENTED ERAGROSTIS, AND CREEPING ERAGROSTIS

When the warm colour of Bent-grasses has faded, these grasses of late summer intensify, with deep violet and purple, the gold of harvest.

One of the most common species, Purple Eragrostis, called by children "Tickle-grass," grows in low tufts on dry and sandy soil, where the gauzy flowering-heads, a foot long or more, spread above the dark green, hairy leaves As the sunlight of early morning falls

"Across the meadows laced with threaded dew"

the flowering-heads of this grass glisten with an intense colour which is reflected in each crystal dewdrop that gems the spikelets In dry fields, where the September sun has burned to a golden brown the shorter growth of grasses, ripening panicles of Purple

ANDERA *Poaceae*

TUFTED LOVE GRASS

Eragrostis, like a reddish purple mist, often cover the ground, and although October frosts fade the flowering-heads to a pale straw-colour, they are still noticeable during autumn, when, as one of the tumbleweeds of the East, they are carried by the wind and piled in huge drifts against wayside fences.

In similar locations the panicles of Lace-grass (*Eragróstis capilláris*) in green and purple are sometimes mistaken for those of the larger species but should be distinguished by the shorter, few-flowered spikelets, and by the absence of hairs surrounding the base of the branches. In both these grasses the widely spreading panicle is usually much longer than the stem which supports it, and the hair-like pedicels are as long or longer than their spikelets.

Tufted Eragrostis (*Eragróstis pilòsa*) is a slender annual which is found by waysides and on sandy river banks. The leaves are very narrow, and the green or purple panicles are shorter and narrower than in the two species mentioned above.

The ornamental grasses of old-time gardens are called to mind as the Strong-scented Eragrostis opens its showy panicles, though surely this plant never found a place by beds of lavender and rosemary, for it emits a most offensive odour, which happily is not possessed by other grasses of the Eastern States. The panicles are not long, being rarely more than six inches in length, but they are closely flowered with large,

Strong-scented Eragrostis
Eragrostis megastachya

179

Eragrostis pectinacea

showy spikelets which in bloom make the grass the most stately of the genus.

Creeping Eragrostis (*Eragróstis hypnoìdes*), found on the sandy banks of streams, spreads extensively over the surface of the ground, and in this habit is unlike our other common species of Eragrostis. The short, dark panicles are somewhat like those of the Strong-scented but are smaller and less densely flowered, and a careful examination shows that the staminate and pistillate flowers are borne on separate plants.

The plants of this genus are widely distributed over the warmer regions of the world, and although these grasses are of less value than many, a large species, found in the Eastern Hemisphere, furnishes an edible grain.

The English gave the name of "Love-grass" to certain of the genus, and it has been suggested that the generic name of these beautiful grasses may have been derived from *eros*, love, and *agrostis*, grass.

Strong-scented Eragrostis. Snake Grass. *Eragróstis megastàchya* (Koeler) Link

Annual. Naturalized from Europe.
Stem 6'-30' tall, erect or spreading, usually much branched. Sheaths

usually smooth, hairy at throat   Ligule a ring of short hairs. Leaves 2'-8' long, 1"-3" wide, flat, rough on margins

Panicle 2'-8' long, densely flowered with large spikelets. Spikelets 8-40-flowered, 2½"-8" long, flat.   Outer scales nearly equal, acute; flowering scales obtuse, 3-nerved.   Stamens 2 or 3   Grass unpleasantly scented

Cultivated lands and waste places   July to September

Throughout nearly the entire United States, and in Ontario

Purple Eragrostis. *Eragróstis pectinácea* (Michx.) Steud.

Perennial, tufted

Stem 1-3 ft tall, erect or spreading. Sheaths smooth or hairy   Ligule a ring of hairs. Leaves 4'-12' long, 1"-4" wide, smooth on lower surface, rough above, hairy near base

Panicle 6'-20' long, pyramidal, reddish purple, the branches 2'-8' long, widely spreading, bearded with white hairs in the axils   Spikelets 4-12-flowered, flat, 1½"-4" long, on pedicels as long or longer   Outer scales acute, about equal, flowering scales acute, 3-nerved, small. Stamens 2 or 3, anthers purple.

Dry soil. July to September

Massachusetts to South Dakota and Colorado, south to Florida and Texas.

NARROW MELIC-GRASS AND PURPLE OAT

"Farre away I heard her song,
'Cusha! Cusha!' all along,
Where the reedy Lindis floweth,
    Floweth, floweth,
From the meads where melick groweth
Faintly came her milking song"

This beautiful grass of spring and early summer, the Narrow Melic-grass, is found by the borders of thickets and open woods from Pennsylvania southward, where its pale green flowers often nod to the breeze above purplish blue banks of dwarf iris and the more fragrant, but somewhat less common, crested iris   So widely open are the dropping spikelets that the panicles seem fringed with pendent green bells, for the papery outer scales of each spikelet are large and broad, like flower petals

A more northern species belonging to this genus is so like the Oat-grasses in appearance that it is commonly called Purple Oat, and in older botanies is given as *Avèna striáta*   The scales are narrow, instead of broad as in the preceding species, and the flowering scale bears an awn as long as itself. Blooming in mid-summer, this grass prefers the dry soil of rocky hillsides. where fragrant pennyroyal and life-everlasting grow in the bor-derland surrounding woods of oak and chestnut. The stems are slender, and the loosely flowered panicles of narrow, long spikelets are usually tinged with purple.

Narrow Melic-grass. *Mé-lica mùtica* Walt.

Perennial.

Stem 1½-3 ft. tall, slender, erect. Sheaths rough. Ligule 1″-2″ long. Leaves 3′-9′ long, 1″-5″ wide, flat, roughish.

Panicle 3′-10′ long, narrow, branches few, not many-flow-ered. Spikelets 3″-5″ long, nodding on slender pedicels and usually consisting of 2 perfect flowers. Rachilla pro-longed and bearing 2 or 3 small, twisted scales. Outer scales slightly unequal, very broad, acute or obtuse; flowering scales papery, broad, obtuse, roughish. Stamens 3.

Rich soil and open woods. April to June.

Pennsylvania to Wisconsin, south to Florida and Texas.

Purple Oat. *Mélica striàta* (Michx.) Hitchc.

Perennial.

Stem 1-2 ft. tall, slender, erect. Ligule very short. Leaves 1′-7′ long, 1″-3″ wide.

Panicle 2′-6′ long, few-flowered, branches slender. Spikelets 3-6-flowered, 8″-12″ long, usually purple. Outer scales unequal,

Narrow Melic-grass
Melica mutica

Purple Oat
Melica striata

182

acute; flowering scales short-hairy at base, divided at apex and bearing a dorsal awn about 4″-5″ long. Stamens 3.

In woods and on rocky hills in the shade. June to August.

New Brunswick to British Columbia, south to Pennsylvania.

BROAD-LEAVED SPIKE-GRASS, SLENDER SPIKE-GRASS, AND SEASIDE OATS

"by rushy brook,
Or on the beachèd margent of the sea."

So strikingly ornamental are the panicles of Broad-leaved Spike-grass that one assumes it to have been among the cherished plants removed from English homes and carefully cultivated in the walled gardens of long ago. But instead this is distinctly an American grass, as are the several members of the genus. Blossoming in late summer, in the borders of moist woods and along winding streams, the large panicles call to mind those old-time bouquets of dried grasses that needed but a touch to start them trembling with faint intimations of the music that was theirs when the breeze, passing over the flowering-heads, shook the ripening spikelets, one against another. The rich green leaves are numerous and widely spreading, and above them the panicles rise like those of a giant Brome-grass, the long, slender pedicels drooping with the weight of the broad spikelets.

Slender Spike-grass (*Uniola láxa*) blooms at the same season in sandy soil, usually near the coast, but shows little resemblance

Broad-leaved Spike-grass
Uniola latifolia

183

to its larger relative. The leaves are long and narrow, and the slender, wand-like panicles bear wedge-shaped spikelets which are nearly sessile and are rarely more than one quarter of an inch in length.

Seaside Oats (*Unìola paniculàta*) is a southern beach grass, growing in drifting sands from Chesapeake Bay southward, and taking the place of the Marram Grass of more northern coasts. Like the Broad-leaved Spike-grass the panicles of Seaside Oats are large and ornamental, but the pale spikelets are more numerous and are borne on much shorter pedicels than are those of the inland species. The stems are stout and erect, with long, narrow, slender-pointed leaves which become tightly rolled as the grass ripens. This species blooms in late fall and often retains the showy blossoming heads through the winter months.

Broad-leaved Spike-grass. *Unìola latifòlia* Michx.

Perennial.

Stem 2-5 ft. tall, stout, erect. Ligule very short. Leaves 4'-10' long, 4"-12" wide, flat, rough on margins, usually hairy at base.

Panicle 5'-12' long, branches slender, drooping. Spikelets broad, many-flowered, 9"-15" long, flat, on hair-like, drooping pedicels. Outer scales slightly unequal, acute, much smaller than flowering scales; flowering scales acute, rough-hairy on their winged keels. Stamen 1.

Moist, shaded places. July to September.

Pennsylvania to Florida, west to Kansas and Texas.

MARSH SPIKE-GRASS

On the salt marshes of midsummer grows a grass whose staminate and pistillate flowers are borne on separate stems, and, although plants of the two sexes are often scattered over the same ground, an acre or more is sometimes covered by stamen-bearing plants, while not far distant an

Marsh Spike-grass
Distichlis spicata

184

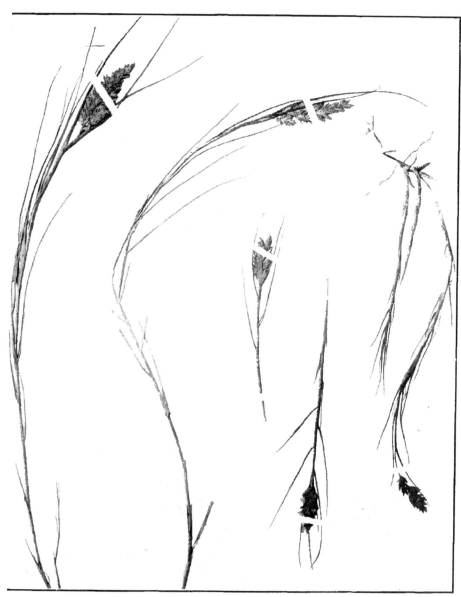

MARSH SPIKE-GRASS (*Distichlis spicata*)

area of like extent shows only pistillate spikes   Marsh Spike-grass, or Salt-grass, as it is sometimes called, is one of the sand-binding grasses, spreading by strong rootstocks, and thriving even on the alkaline deserts of the interior where there is little vegetation, and where the presence of this grass is welcomed by thirsty travellers as a certain indication of water near the surface of the soil

A tough wiry grass it is, like a denizen of inhospitable ground; the low stems, erect and rigid, bear stiff leaves and short, compact, spike-like panicles of straw-coloured blossoms. Upon new land the straight rootstocks, according to Mr. Coville, send up their erect stems at intervals of about four inches, and until the grass is fully established these stems appear to cut the ground into triangles, quadrangles, and similar geometrical figures.

Marsh Spike-grass.   Salt-grass. *Distichlis spicàta* (L.) Greene

Perennial, from creeping rootstocks

Stem 6'-24' tall, wiry, erect. Ligule a ring of short hairs   Leaves $\frac{1}{2}'$-6' long, 1"-2" wide, rather rigid, flat or involute

Spike-like Panicle 1'-2$\frac{1}{2}$' long, densely flowered   Staminate and pistillate flowers borne on separate plants   Spikelets 4-18-flowered, 4"-9" long, yellowish green, more numerous on staminate plants   Outer scales acute, unequal, flowering scales acute, broader and longer than empty scales. Stamens 3   Stigmas long

Salt marshes and saline soils   June to August

Maine to Florida and Texas, also on the Pacific coast, and in alkaline soil in the interior

LADY'S HAIR, OR QUAKING-GRASS

Lady's Hair, Lady's Mantle, Lady's Shoes — to continue the list no further — are examples of wayside plants in which a devout people saw articles of person and attire belonging to the Blessed Virgin. As patroness of those flowers which are dedicated to her under the name of "Lady" the Virgin has an ever-living wardrobe, including even her nightcap, and furnishing purse and thimble, though for possession of her comb she has to dispute, not only with the ever-beautiful goddess but also with that personage who is reported as going to and fro in the earth, and walking up and down in it, and who, with scant use for toilet articles, was evidently thought by our ancestors to have required the same plant for darning needles!

187

Lady's Hair, or Quaking-grass, is not often seen in American fields, yet it has become sparingly naturalized in the Eastern States

where it has escaped from that cultivation as an ornamental grass for which it was brought from Europe long ago. This grass is a slender perennial, blooming in late spring, and bearing numerous inflated, heart-shaped spikelets of lavender and green, which droop on pedicels so slender that the slightest breath causes them to tremble.

In the old Doctrine of Signatures, which saw more things in heaven and earth than philosophy dared dream, that which shook was a panacea for diseases of trembling, and in older days, in many counties of England, Quaking-grass was gathered to ensure freedom from ague.

Lady's Hair. Quaking-grass. Shaking-grass *Brìza mèdia* L.

Perennial. Naturalized from Europe.

Stem 6'-24' tall, slender, erect. Ligule very short. Leaves 1'-4' long, 1''-2½'' wide, flat.

Panicle 1½'-5' long, pyramidal, open, branches slender, not numerous. Spikelets 5-12-flowered, 2''-3'' long, purplish, inflated, ovate or heart-shaped, borne on drooping pedicels. Outer scales nearly equal, broad, concave; flowering scales concave, broader than outer scales; palets much shorter than flowering scales. Stamens 3.

Fields and waste places. May to July.

Ontario to southern New England.

ORCHARD GRASS

Spreading tufts of the blue-green leaves of Orchard Grass are very noticeable by the waysides of early spring, before the pageant of summer brings a score of grasses to every lane and byway. In many states this is one of the most common species, and is

Lady's Hair
Brìza media

THE FIELD IN WHICH ARE FOUND SWEET VERNAL-GRASS, ORCHARD-GRASS, TIMOTHY, KENTUCKY BLUE-GRASS, AND MEADOW FESCUE

the first of the larger grasses to bloom. The stout stems grow rapidly, and when clover fields are sweet with blossoms the coarse panicles of Orchard Grass are painted with large anthers of purple and yellow, terra-cotta and pink, the colour varying with the soil and the abundance of light. The few branches of the flowering-head spread stiffly, and near their extremities the spikelets are crowded in dense, one-sided clusters.

Orchard Grass is one of the most widely known of cultivated grasses, and is one that is highly valued by the farmer, since the rank growth, both in the pasture and as aftermath in the field, makes it for him the earliest grass in spring and the latest in autumn. It grows especially well in shaded places, where few grasses attain luxuriant growth, and in old orchards the coarse tussocks are very common. The sheaths differ from those of the majority of grasses in that they are perfectly closed until the inflorescence, forcing its way up, causes them to split.

Like many grasses that were brought from Europe at an early date, Orchard Grass attracted little attention in England until re-introduced to that country

Orchard Grass

Dactylis glomerata

212

from America. The English name of Cock's-foot Grass is derived from a fancied resemblance between the branching panicle and a bird's foot.

Orchard Grass. Cock's-foot Grass. *Dáctylis glomeráta* L.

Perennial, tufted. Naturalized from Europe.

Stem 2-5 ft. tall, coarse, erect. Ligule 1"-3" long. Leaves 4'-14' long, rough, flat or slightly keeled.

Panicle 3'-9' long, branches coarse, rough, widely spreading in flower. Spikelets 3-5-flowered, 3"-4" long, green or purple, in dense 1-sided clusters at the ends of the branches. Outer scales unequal, keeled, sharply pointed; flowering scales awn-pointed, rough. Stamens 3, anthers yellow, terra-cotta, pink, or purple.

Fields, waysides, and dooryards. May to July.

New Brunswick to Manitoba, south to South Carolina, Kansas, and Colorado.

CRESTED DOG'S-TAIL

The rough, narrow, spike-like panicles of Crested Dog's-tail are seldom found save in waste places and by waysides, since this grass, as yet, has hardly become naturalized in America. It is a slender species and differs from our common grasses in that it bears both sterile and perfect spikelets which are arranged in clusters. In the perfect flowers the scales are much broader than are the rough scales of the sterile spikelets.

As the roots of the Crested Dog's-tail penetrate deeply into the earth the leaves remain fresh and green when other grasses are partially withered, and Sinclair, who carried on extensive researches in the study of English grasses, found this species to yield a large part of the herbage of the most celebrated pastures he examined in that country.

The grass blooms in midsummer, and so fine and strong are the slender stems that in foreign countries, when material for straw-plaiting is gathered, taller grasses are passed by for this, which is said to be much used in the making of Leghorn hats.

Crested Dog's-tail
Cynosurus cristatus

194

Crested Dog's-tail. Dog's-tail Grass. *Cynosùrus cristàtus* L.

Perennial   Introduced from Europe.
Stem 1-2½ ft tall, slender, erect.   Ligule very short   Leaves 1'-5' long,
  ½"-2" wide, flat
Spike-like Panicle 2'-4' long, narrow   Spikelets of two kinds in small
  clusters, lower spikelets of the clusters larger, consisting of several
  or many rough, narrow, empty scales, upper spikelets consisting of a
  few sharp-pointed, broader scales enclosing perfect flowers, flowering
  scales about 1½" long   Stamens 3.
Fields, waysides, and waste places   June to August.
Newfoundland to Ontario, south to New Jersey.

THE POAS

LOW SPEAR-GRASS, KENTUCKY BLUE-GRASS, ROUGH-STALKED
MEADOW-GRASS, WOOD SPEAR-GRASS, CANADA BLUE-
GRASS, AND FALSE RED-TOP

These meadow grasses should be listed among the plants that
give us

"The flower of every valley, the flower of all the year,"

since from early spring until late autumn some representative
of the genus may be found in bloom, and in the most Southern
States there is rarely a month when Low Spear-Grass is not in
flower. In the Northern States this species is one of the earliest
plants to change the brown hillsides to living green, and on lawns
and by waysides small tufts of this modest little grass are common
throughout nearly the whole country. Even between the flag-
stones of the city, Low Spear-grass tries to obtain foothold, often
succeeding and blossoming, though choked by dust and daily
trodden under foot.   The flattened stems, usually but six or eight
inches in height, bear short, yellowish green panicles which, unlike the flowers of other early grasses, are sent up during the
entire summer
  Closely following the blossoming of Sweet Vernal-grass the
famous Kentucky Blue-grass adds the delicacy of its graceful
panicles to the common garden of the wayside   Although well
known, by name at least, few seem acquainted with the fact that
this is one of our most common grasses from the Atlantic to the
Pacific   Its most luxuriant growth is attained in the far-noted
blue-grass region of Kentucky, on limestone soils in the counties

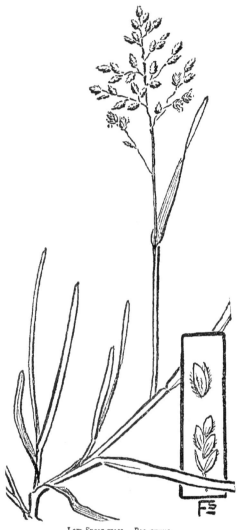

Low Spear-grass. *Poa annua*

about Lexington, that "city of the Blue-grass," but even in the more eastern states the slender stems are seen as frequently as are those of any other grass of early summer, and the profusion of dark green leaves, together with the habit of spreading by sending off numerous running rootstocks, renders it an ideal turf-forming grass. June Grass, as this species is often called, is really a more appropriate name than Blue-grass, as the plants lack the deep blue-green colour which characterizes Canada Blue-grass, the true "blue-grass" of the genus. The blossoming head, in varying shades of green, lavender, and purple, is in form a perfect pyramid, and as the flowers bloom in June, before the summer grasses, the plants should be easily recognized. In dry or sandy soil the grass is small and harsh, but in richer grounds the stems are from two to four feet tall, and later in the season, when the green has faded, they stand like threads of shining gold by every wayside.

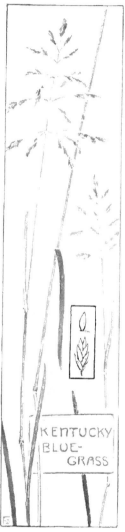

KENTUCKY
BLUE-
GRASS

Poa pratensis

Rough-stalked Meadow-grass (*Pòa trivìàlis*) resembles Kentucky Blue-grass, but is less common, and may be distinguished by its long ligule and rough sheaths. It is usually more slender than the preceding species, and it does not spread by rootstocks.

Wood Spear-grass (*Pòa alsòdes*), a slender grass of wooded hillsides, blooms in May and June and shows narrow, rather loose panicles of small green spikelets. The sheaths are longer than the internodes, and the upper sheath frequently encloses the base of the panicle.

In early spring the leafy shoots of Canada Blue-grass (*Pòa compréssa*), the

107

bluest of the Poas, are noticeable in every soil; on sandy hills and in the thickets that border deep woods, on scantily covered rocks and by trodden paths. The whitish summits of the sheaths are very conspicuous against the blue-green leaves, and although the plants vary greatly in size they are rarely more than two feet tall, and are constant in their character-istic colour and in the strongly flattened stems. Unlike the Kentucky Blue-grass, which soon ripens, the Canada Blue-grass blooms the entire season. Its panicles are short and narrow (usually one-sided), with s h o r t branches and greenish spike-lets.

False Red-top, the tall-est of the common Poas, blooms in swampy places and in wet meadows, where the green spikelets show each a tawny orange tip and sometimes change to dull purple as the seeds ripen. The large, gracefully droop-ing panicles could hardly be mistaken for those of the Red-top of the fields, and assuredly not if the spikelets were exam-ined, showing several tiny flowers in each spikelet where the Red-top has but one. In many places this species is known as Fowl Meadow-grass, and the tradition is that it received that name from the fact that wild ducks and other water-fowl brought the seed to a low meadow near Dedham, Mass.

Flexuous Spear-grass and Short-leaved Spear-grass (*Pòa autumnàlis* and *P. brachy-*

False Red-top, or Fowl Meadow-grass. *Poa triflora*

CANADA BLUE-GRASS (*Poa compressa*).   One half natural size

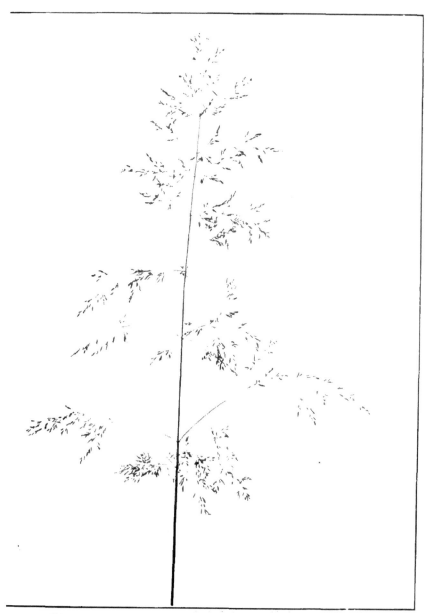

FALSE RED-TOP OR FOWL MEADOW GRASS

WOOD SPEAR-GRASS (*Poa sylvestris*).   Natural size

phylla) are the earliest of the genus and are found in woods from New York State southward. They bloom in March, when the first arbutus opens, and are slender grasses, with loose panicles which bear but a few green spikelets at the extremities of the branches.

Each flowering scale in the majority of the species of this genus shows a small tuft of cobwebby hairs at the base, and under the microscope this tuft forms a distinguishing feature by which the grasses may be recognized. In Canada Blue-grass and Flexuous Spear-grass the tufts are lacking, but the flowering scales are downy below the middle.

Low Spear-grass. Dwarf Meadow-grass. *Pòa ánnua* L

Root annual. Naturalized from Europe

Stem 2'-12' tall, erect or spreading, flattened   Sheaths loose   Ligule about 1" long. Leaves ½'-4' long, about 1" wide, flat

Panicle ½'-4' long, pyramidal, open, often 1-sided, branches short   Spikelets 3-6-flowered, about 2" long   Outer scales slightly unequal, 1st scale acute, 2nd scale obtuse, flowering scales obtuse, hairy at the base   Stamens 3

Fields, waysides, and cultivated grounds   April to October

Throughout nearly the whole of North America.

Kentucky Blue-grass. June Grass. *Pòa praténsis* L.

Perennial, with rootstocks.   Naturalized in the Eastern States, indigenous elsewhere

Stem 8'-4 ft. tall, slender, erect.   Ligule very short   Leaves 1"-3" wide, flat, stem leaves 2'-6' long, basal leaves much longer.

Panicle 2'-8' long, pyramidal, open, branches slender, lower branches ½'-3' long   Spikelets 3-5-flowered, about 2" long, green or purplish   Outer scales acute, unequal, roughish on keels, flowering scales acute, webby at base, downy below on marginal nerves and mid-nerve; palets nearly as long as flowering scales.   Stamens 3, anthers often deep purple

Waysides, fields, and meadows   May to August.

Throughout nearly the whole of North America

False Red-Top. Fowl Meadow-grass. *Pòa triflòra* Gilib

Perennial.

Stem 2-5 ft. tall, erect, rather slender   Ligule 1"-2" long   Leaves 4'-10' long, 1"-2" wide, flat

Panicle 6'-12' long, branches rough, slender, divided and spikelet-bearing above the middle, lower branches 2'-5' long   Spikelets 2-5-flowered, 1½"-2" long, on short pedicels   Outer scales acute, slightly unequal, flowering scales obtuse, webby at base, downy below on marginal

nerves and mid-nerve, usually tawny orange or reddish at apex;
palets nearly as long as flowering scales   Stamens 3.
Wet meadows and swampy places   June to August
Nova Scotia to Vancouver Island, south to New Jersey, Illinois, and
Nebraska.

THE MANNA–GRASSES

NERVED MANNA-GRASS, TALL MANNA-GRASS, RATTLESNAKE GRASS,
DENSELY FLOWERED MANNA-GRASS, FLOATING MANNA-GRASS,
AND SHARP-SCALED MANNA-GRASS

In June, when the low swales by the brooks are full of interest,
and a score of flowering plants may be gathered from the vantage
ground of a drier tussock in the marsh, the graceful Manna-grasses
cover large areas and bloom in tones of dull green and purple,
darker where the sun has burned them longer, but typical of spring
as are the nearby orchids, and ever lacking that suggestion of midsummer heat which the reddish purple Bent-grasses bring to July
fields

Nerved Manna-grass usually precedes the other species by a
fortnight, and is perhaps the most common in a majority of the
states.   Growing luxuriantly in the borderland between pasture
and marsh it furnishes an important part of the herbage of wet
meadows, and, though it varies greatly in different soils, the
gracefully drooping panicles may be recognized by their spreading and drooping branches and their tiny, purple and green
spikelets

Tall Manna-grass (*Glycèria grándis*), a stout, handsome species,
is often seen in wet grounds, where the ample panicles and broad
leaves rise above sedges and low grasses   Like others of the
genus, Tall Manna-grass is a species of which cattle are fond and
wade through miry bogs to reach, and water fowl, during the fall
migration, find resting places along streams where these grasses
grow abundantly, as the seeds yield a feast to thousands of birds.
Tall Manna-grass is from three to five feet in height and bears
large panicles of many spikelets.

Heavy, drooping panicles of Rattlesnake Grass are found lean-
ing over narrow brooks and ditches, and by damp waysides where
meadow-rue and sedges luxuriate   With pendent, inflated spikelets of pale green and purple this grass is the most beautiful of the

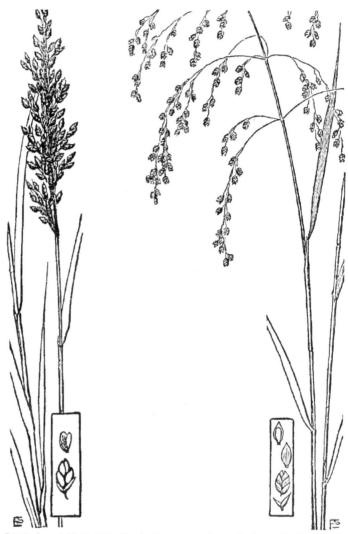

Densely flowered Manna-grass. *Glyceria obtusa*   Rattlesnake Grass. *Glyceria canadensis*

207

genus, and is in its greatest perfection in late June, when the low grounds are

"paynted all with
variable flowers,
And all the meades
adorned with daintie
gemmes."

The flowering-heads of this grass retain their beauty until late fall, and are easily recognized throughout the season, though the colours which tinge the broad scales fade as the seeds ripen.

Densely flowered Manna-grass is less widely distributed, and in bloom is quite unlike other marsh grasses. It should be recognized by the erect, bunch-like inflorescence of crowded spikelets.

Floating Manna-grass is often found in shallow, running water, but the long panicles bear little resemblance to the flowering-heads of Nerved Manna-grass or Rattlesnake Grass. The spikelets of Floating Manna-grass are long and narrow, and the branches, at first closely appressed, at last spread rather stiffly from the stem. The manna crop of Germany and Poland is gathered from a species similar to this, and the seeds are there considered desirable in soups and gruels. Bread made from the meal is said to be very little inferior to that made from wheat, but the American farmer has little time to experiment with so small a grain when the product of years of cultivation is ready for his sowing, and in this country birds gather the harvest by the water's edge, while, as the tall stems lean over streams, the fallen seeds are eagerly eaten by fish.

Sharp-scaled Manna-grass (*Glycèria acutiflòra*) is a pale green

Nerved Manna-grass
Glyceria nervata

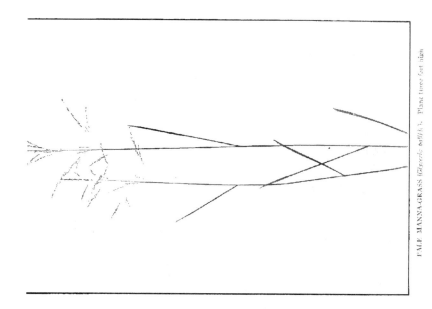

FALSE MANNA-GRASS (*Glyceria pallida*). Plant three feet high

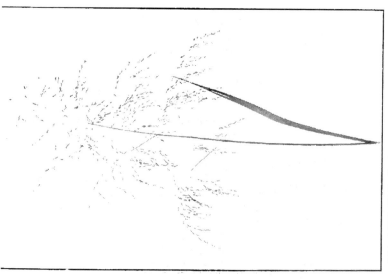

TALL MANNA-GRASS (*Glyceria grandis*). Panicle one-half natural size

NERVED MANNA GRASS—(*Glyceria nervata*). One-half natural size.

grass, recognized by the long, narrow spikelets which protrude through the enclosing sheaths as the plant begins to bloom. This grass has a slight resemblance to Floating Manna-grass, but is much smaller, though the spikelets are longer, being from one to one and three quarters inches in length. The panicles are long and narrow, with short, erect branches, and the acute flowering scales are shorter than the long-pointed palets.

Rattlesnake Grass. *Glycèria canadénsis* (Michx.) Trin.

Perennial.

Stem 2-3 ft. tall, erect. Ligule about 1" long. Leaves 6'-15' long, 2"-4" wide, rough, flat, spreading at right angles to stem.

Panicle 5'-10' long, nodding, branches rough, spreading or drooping, lower branches 2'-6' long. Spikelets 5-12-flowered, 2½"-4" long, broad, inflated, flattened, green tinged with purple. Outer scales acute, unequal, shorter than flowering scales; flowering scales broad, obtuse or acute, obscurely 7-nerved; palets broad, slightly shorter than flowering scales. Stamens commonly 2.

Wet meadows, brooksides, marshes, and swamps. June to August.

Newfoundland to Ontario and Minnesota, south to New Jersey and Kansas.

Densely flowered Manna-grass. *Glycèria obtùsa* (Muhl.) Trin.

Perennial.

Stem 1-3 ft. tall, stout, erect. Ligule very short. Leaves 6'-15' long, 2"-4" wide, flat, dark green.

Panicle 3'-8' long, densely flowered, contracted, branches erect. Spikelets 3-7-flowered, 2"-3" long, somewhat inflated. Outer scales acute, unequal, shorter than flowering scales; flowering scales broad, obtuse, obscurely 7-nerved; palets slightly shorter than flowering scales. Stamens 2 or 3.

Swamps and wet places. July to September.

New Brunswick to New York, south to North Carolina.

Floating Manna-grass
Glyceria septentrionalis

213

Nerved Manna-grass. *Glycèria nervàta* (Willd.) Trin.

Perennial.

Stem 1-3½ ft. tall, slender, erect. Ligule 1″-2″ long. Leaves 6′-12′ long, 2″-5″ wide, flat, smooth on lower surface, rough above.

Panicle 3′-10′ long, pyramidal, open, somewhat nodding, branches slender, spreading or drooping, rough, lower branches 2′-6′ long. Spikelets numerous, 3-7-flowered, 1″-2″ long, dark green tinged with purple. Outer scales obtuse, unequal, shorter than flowering scales; flowering scales obtuse, sharply 7-nerved; palets as long as flowering scale. Stamens 2 or 3.

Wet meadows, brooksides, and marshes. June to September.

Newfoundland to British Columbia, south to Florida and Mexico.

Floating Manna-grass. *Glycèria septentrionàlis* Hitchc.

Perennial.

Stem 2-5 ft. tall, rather stout, somewhat flattened, erect or spreading at base. Sheaths loose. Ligule 2″-3″ long. Leaves 6′-15′ long, 2″-6″ wide, flat, roughish.

Panicle 8′-14′ long, branches appressed, finally spreading, not numerous, lower branches 3′-6′ long. Spikelets 7-13-flowered, 4″-12″ long, narrow, green, appressed on the branches. Outer scales unequal; 1st scale obtuse or acute; 2nd scale obtuse; flowering scale obtuse, roughish, 7-nerved; palets slightly longer than flowering scales. Stamens 2 or 3.

Wet places and in shallow water. June to September.

Maine to North Carolina and westward.

GOOSE–GRASS AND SPREADING SPEAR–GRASS

These grasses, which in their spikelets bear a resemblance to the Manna-grasses, are found only on the salt marshes and beaches. They are slender grasses blossoming in midsummer and later when the beautiful sea-pink, or rose of Plymouth, blooms in salt meadows, and the marsh rosemary, growing to the water's edge, colours the sand with a mist of lavender flowers, like purple spray borne inland on the waves.

Goose-grass
Puccinellia maritima

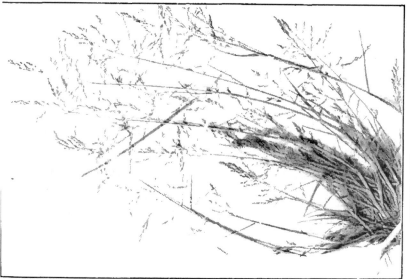

GOOSE-GRASS (*Puccinellia maritima*).   Not common south of Massachusetts

SPREADING SPEAR-GRASS (*Puccinellia distans*)

From Spreading Spear-grass (*Puccinéllia dístans*) Goose-grass differs in that it rises from rootstocks and bears narrow panicles of long spikelets. Spreading Spear-grass, on the other hand, is tufted, without rootstocks, and opens wide panicles of small, crowded spikelets. The latter species is slightly stouter than Goose-grass, and bears wider leaves.

Goose-grass. Sea Spear-grass. *Puccinéllia marítima* (Huds.) Parl.

Perennial, from rootstocks.

Stem 8′-24′ tall, slender, erect or spreading at base. Ligule short. Leaves ½′-5′ long, 1″ wide or less, flat or involute.

Panicle 2′-6′ long, narrow, branches short. Spikelets 3-10-flowered, 3″-6″ long, narrow. Outer scales unequal, obtuse or acute; flowering scales broad, obtuse, nerves very obscure; palets nearly as long as flowering scales. Stamens 3.

Salt marshes and sea beaches. July and August.

Labrador to New Jersey, also on the Pacific Coast.

THE FESCUES

SLENDER FESCUE, SHEEP'S FESCUE, MEADOW FESCUE, RED FESCUE, AND NODDING FESCUE

"The grass-blade, like a long green ribbon, streams from the sod into the summer, checked indeed by the frost, but anon pushing on again, lifting its spear of last year's hay with the fresh life below."

When the beauty of the earliest grasses is passing the Fescues appear, and in the calendar of the grasses mark the beginning of summer. For a short time they take the primary position in the fields. Slender Fescue and Sheep's Fescue are among the first to bloom, and closely following the opening of their flowers the graceful stems of Meadow Fescue, abundant in pastures and meadows, are seen also by the wayside bending over ripening spikes of Sweet Vernal-grass.

Sheep's Fescue (*Festùca ovìna*) and Slender

Slender Fescue
Festuca octoflora

217

Fescue, whose narrow panicles rise above tufts of bristle-like gray-green leaves, are the smallest of all, and are usually found in dry locations. The latter species is distinguished by slightly longer flowering-heads, more numerously flowered spikelets, and longer awns. The stems of both these grasses are usually about a foot in height, or in sterile soil they are often much smaller, though one occasionally finds a tall variety of Sheep's Fescue which bears a more open panicle and larger spikelets.

Red Fescue is locally common by waysides and is found in the shade as well as in the sunlight. Like the two preceding species it has a profusion of involute basal leaves, but unlike them it springs from extensively creeping rootstocks and so is one of the useful soil-binders on drier slopes. This species is variable and is perhaps most easily recognized by the tufts of bristle-like, dark leaves which surround the base of the stems.

The most common of the genus is the Meadow Fescue, which was introduced from Europe many years ago. For so tall a grass the smooth stems are quite slender, and with their tapering, shining leaves are a wide contrast to Timothy, which begins to bloom before the Meadow Fescue has faded, and is so often associated with it in the fields. The long spikelets of Meadow Fescue are green, frequently tinged with reddish purple, and in bloom the flowers for a short time are broadly open, giving delicacy to the one-sided, drooping panicle, which after flowering is narrow and closely contracted.

Rocky woodlands in nearly all the states shelter the Nodding Fescue (*Festùca nùtans*),

Red Fescue
Festuca rubra

218

MEADOW FESCUE.

a slender, dark green grass with loose, few-flowered panicles. The spikelets are small, and as they are borne only at the ends of the panicle branches the plant should not be confused with other shade-loving grasses.

The plants of this genus are very variable under different conditions of soil and climate, and a number of varieties are listed under the species given.

Slender Fescue. *Festùca octoflòra* Walt.

Root annual, often tufted.

Stem 4'-20' tall, slender, erect, wiry. Ligule very short. Leaves bristle-like, 1'-3' long, involute, occasionally downy.

Panicle 1'-6' long, narrow, contracted, often 1-sided, branches short. Spikelets 5-13-flowered, 3"-5" long. Outer scales very acute, slightly unequal, smooth; flowering scales rough, bearing a terminal awn 1"-3" long; palets nearly as long as flowering scales. Stamens 2.

Dry sterile soil. May to August.

New Brunswick to Florida, west to Washington and California.

Red Fescue. *Festùca rùbra* L.

Perennial, with creeping rootstocks.

Stem 1-2½ ft. tall, slender, erect. Ligule very short. Leaves of sterile shoots involute, bristle-like, 3'-10' long, stem leaves shorter, involute in drying, minutely downy on upper surface.

Panicle 2'-6' long, branches ascending, spreading in flower, not numerous, lower branches 1'-3' long. Spikelets 3-9-flowered, 3"-5" long, green or reddish. Outer scales acute, unequal; flowering scales bearing each a short terminal awn; palets as long as flowering scales. Stamens 3, anthers yellow or purplish.

Dry soil. June to August.

Labrador to Alaska, south to North Carolina and Colorado.

Meadow Fescue
Festuca elatior

Meadow Fescue. Tall Fescue. *Festùca elàtior* L.

Perennial. Naturalized from Europe.

Stem 2-5 ft. tall, erect.
Ligule very short.
Leaves 3'-15' long, 2"-4" wide, flat, often rough.

Panicle 4'-12' long, narrow, usually nodding at top, branches spreading in flower, erect before and after blossoming. Spikelets 4-9-flowered, 4"-6" long, green or tinged with purple. Outer scales acute, unequal; flowering scales acute or short-pointed; palets nearly as long as flower-scales. Stamens 3, anthers reddish purple or yellow.

Meadows, fields, and waysides. June to August.
Nova Scotia to Ontario and southward.

THE BROME-GRASSES

DOWNY BROME-GRASS, FRINGED BROME-GRASS, CHESS, AND UPRIGHT CHESS

Many of the Brome-grasses, as emigrants from Europe, have become weeds in this country, and from May until August are found blooming by waysides and in waste places, where the beautiful panicles of large drooping spikelets should be quickly recognized.

Downy Brome-grass, the earliest species, is common along railway embankments and by dry roadsides and is resembled by closely related species which have a more southern range. The plant is a low, slender annual whose panicles of awned, drooping spikelets resemble the heavier heads of cultivated oats. The stems are reddish near the nodes and soon turn to shining purple, while the ripening flowering-head is tinged with the same colour. In sandy locations the whole plant dries in a few weeks, and, faded to a pale

Fringed Brome-grass
Bromus ciliatus

222

FUNGID BROMEGRASS

BROME GR... ward

straw-colour, remains throughout the season, the heads bristly and lacking the gracefulness that was theirs in early spring.

Fringed Brome-grass, one of the native species, is most frequently found in low grounds, where meadow and woodland meet in a debatable border of half thicket, half marsh, as the meadow grasses give place to sedges and a few stragglers from the thickets advance toward more open country. The stems of Fringed Brome-grass are stout and leafy, usually rising in groups which are very noticeable above a lower growth of plants. The panicles are large and are composed of slender branches bearing silky, short-awned spikelets.

Handsome groups of Chess are frequently seen in old grain fields and on waste land, where this grass appears as a weed, and in midsummer opens heavy panicles of large spikelets. If every plant is sometime "to be of utility in the arts" Chess has as yet shown nothing but beauty as its excuse for appearing so often where it is least wanted. The panicles are striking and ornamental, but Chess has met little favour either in this country or abroad. With gifted imagination, and untroubled by the constancy of Nature, the peasantry of the Old World considered this grass a degenerated wheat, and supplied the missing links in the lineage by assuming sundry transmutations in which a grain of wheat should send up a stalk of rye, and the rye being sown should produce barley, while from barley a Chess should be grown that later, under favorable conditions, might awaken to life under the form of oats.

Downy Brome-grass
Bromus tectorum

229

Even the earlier farmers of this country thought this grass a wheat that had fallen to low estate, and so called it "Cheat."

Upright Chess (*Bròmus racemòsus*), also found in grain fields and in waste places, is sometimes mistaken for Chess, from which it differs in the more slender stem, narrower, shorter, and more erect panicles, plainly nerved flowering scales, and shorter palets.

Fringed Brome-grass. Swamp Chess.
Bròmus ciliàtus L.

Perennial.

Stem 2-4 ft. tall, erect, usually stout and leafy. Sheaths closed, split near top, frequently downy. Ligule very short. Leaves 5'-12' long, 2"-6" wide, flat, roughish, usually downy on upper surface, dull or pale green.

Panicle 4'-10' long, branches slender, widely spreading or drooping, lower branches 2'-5' long. Spikelets 5-8-flowered, about 1' long, green. Outer scales acute, unequal, rough on keels; flowering scales 4"-6" long, obtuse or acute, downy near margins, 2-toothed at apex and bearing a short, straight awn 2"-4" long; palets slightly shorter than flowering scales. Stamens 3.

Damp soil in open woods and borders of thickets. June to August.

Newfoundland to New Jersey, west to Manitoba and Minnesota.

Downy Brome-grass. *Bròmus tectòrum* L.

Root annual. Naturalized from Europe.

Stem 6'-24' tall, slender, erect or spreading. Sheaths downy. Ligule 1"-2" long. Leaves 1'-4' long, 1"-2" wide, downy, flat.

Panicle 2'-6' long, branches slender, drooping. Spikelets 5-8 flowered, 6"-12" long, on slender, drooping pedicels. Outer scales acute, unequal, rough-hairy, 2nd scale slightly 2-toothed; flowering scale 3"-6" long, rough or hairy, acute, 2-toothed and bearing from between the teeth a straight awn 5"-8" long. Stamens 3.

Waste places and waysides. May to July.

New England to Illinois and southward.

Chess
Bromus secalinus

Chess. Cheat. *Bròmus secalìnus* L.

Annual. Naturalized from Europe.

Stem 1-4 ft. tall, erect, rather stout. Sheaths usually smooth. Ligule short. Leaves 3'-10' long, 2"-4" wide, flat, somewhat hairy, conspicuously veined.

Panicle 2'-8' long, pyramidal, branches spreading or drooping, lower branches ½'-4' long. Spikelets 6-10-flowered, 6"-10" long. Outer scales unequal; 1st scale acute; 2nd scale obtuse; flowering scales 3"-4" long, obtuse, often downy on upper margins, awnless or bearing a short, straight awn from between the obtuse teeth; palets about as long as flowering scales. Stamens 3.

Fields and waste places, especially in grain fields. June to August.

Nearly throughout North America except in the extreme north.

RAY-GRASS, DARNEL, AND ITALIAN RYE-GRASS

Honour should be granted the Ray-grass, since it was probably the first of the grasses noticed and cultivated as a forage plant. In Europe its use extends over many scores of years, and it is certain that in England Ray-grass has been held in esteem since the days of Charles II, though it was not until after the middle of the eighteenth century that other grasses were considered worth the care of gathering and sowing.

Like many of the field grasses Ray-grass has long been naturalized from Europe, but it is less frequently cultivated in this country than are other grasses more suited to our soil and climate. The slender elongated spikes, rising in midsummer, are beaded with edgewise placed spikelets which open stiffly in flower and are light green in colour with pale pendent anthers.

Ray-grass
Lolium perenne

Closely related to this species is the *"infelix lolium"* of Vergil—

Lòlium temulèntum — supposed by some to have been the "tares among the wheat" mentioned in St. Matthew's Gospel This latter species, the Darnel, an annual occasionally found as a weed in grain fields, is remarkable for the poisonous quality of its seeds which cause serious trouble if the "tares" are gathered with the wheat and the seeds find their way to the mill with the pure grain The most noticeable difference between this grass and Ray-grass, to which the name of Darnel is sometimes erroneously applied, is that in the true Darnel the long outer scale fully equals, and often exceeds, its spikelet in length. In Scotland the name of "Sleep-ies" has been given to Darnel on account of the narcotic effect of its seeds, though more recently it has been said that only the dis-eased, or ergotized, grain is poisonous

Italian Rye-grass (*Lòlium multiflòrum*) has been brought in later years to the United States. From either of the preceding species it is distinguished by its ten to twenty-flowered spikelets.

Ray-grass. Rye-grass. Ray-darnel. *Lòlium perènne* L.

Perennial. Naturalized from Europe.
Stem 1-3 ft tall, erect. Ligule short. Leaves 2'-8' long, 1"-2½" wide, flat, roughish
Spike 3'-9' long, narrow Spikelets 5-12-flowered, 4"-8" long, green, solitary, sessile on alternate notches of the rachis, edge of each spike-let (or backs of the scales) turned toward the rachis Two empty scales in terminal spikelet, only one empty scale in other spikelets Empty scale acute or obtuse, dark green, thick, strongly nerved, flowering scales acute or short-awned, occasionally obtuse, palets nearly as long as flowering scales Stamens 3.
Fields, waysides, and waste grounds June to August.
Canada to North Carolina and Tennessee, also in California and Arizona.

COUCH-GRASS, BEARDED WHEAT-GRASS, AND PURPLE WHEAT-GRASS

In June the Couch-grass suddenly appears by the waysides and as the worst of weeds in cultivated lands; the stout leafy stems and flattened two-sided spikes apparently having sprung up in a night. This species varies greatly in appearance, especially near the seacoast, but it is always unlike other grasses, with the possible exception of Ray-grass from which it is distinguished by the posi-tion of the spikelets, those of Ray-grass being placed edgewise,

RAY-GRASS (*Lolium perenne*). Natural size. Spikelets enlarged by two

or with their backs to the stem, while in Couch-grass the spikelets are closely placed with their sides against the axis of the spike.

Couch-grass grows with the energy of the fabled hydra, and where one of the dark green stems is cut, half a dozen rise to take its place. This grass and the Johnson Grass of the South have the most extensive system of creeping or, more expressively, running rootstocks of any of the inland grasses. The strong, white subterranean stems of Couch-grass form a network and send off innumerable sharp-pointed shoots, which in the garden often pierce roots and tubers and seem to prefer to grow through any permeable object rather than to turn aside. This grass is the worst enemy of the farmer among his cultivated acres, as each breaking of the ground's surface by sharp-edged tools serves only to cut and scatter the roots, each fragment of which, seemingly, "hath in it a Propertie and Spirit, hastily to get up and spread." This quality of the plant suggested to Charles Dudley Warner while spending his "Summer in a Garden" the idea of offering Couch-grass to the clergy as an example of total depravity, yet insatiable ambition seems the chief characteristic of this plant, whose merits are recognized in its tenacity of life through drouth and on sandy soils, as well as in the nutritious hay yielded, while the long rootstocks are valuable in binding the loose soil of railway embankments. On pasture lands of the Northwestern States other species of the genus furnish an important part of the native grasses.

Bearded Wheat-grass (*Agropyron caninum*), less common in the East, is unlike Couch-grass in the absence of rootstocks, in the

Couch-grass
Agropyron repens

occasional downiness of the lower sheaths, and in the long-awned scales.

Purple Wheat-grass (*Agropyron biflòrum*), a mountain species, is of smaller growth, bearing shorter leaves and smaller spikes, the latter usually tinged with purple.

Couch-grass. Quick-grass. Quitch-grass. Devil-grass. Witch-grass. *Agropyron rèpens* (L) Beauv.

Perennial, with running rootstocks   Naturalized from Europe.
Stem 1-4 ft tall, erect   Ligule very short.   Leaves 4'-12' long, 2"-5" wide, flat, smooth on lower surface, rough above.
Spike 2'-8' long, narrow.   Spikelets 3-6-flowered, 4"-10" long, green, solitary, sessile on alternate notches of the rachis, side of each spikelet placed against the rachis   Outer scales acute, or awn-pointed, sometimes obtuse or notched, strongly nerved, about equal, flowering scales acute or short-awned, palets slightly shorter than flowering scales   Stamens 3, anthers large, yellow   A very variable species.
Fields, cultivated ground, and waste places.   June to September
Newfoundland to the Northwest Territory, south to Virginia, Ohio, and Iowa.

BARLEY, SQUIRREL-TAIL GRASS, AND WALL BARLEY

"First rie and then barlie, the champion saies,
Or wheate before barlie, be champion waies
But drink before bread-corn with Middlesex men,
Then laie on more compas, and fallow agen"

Occasionally a few grains of the cultivated barley (*Hórdeum sativum*) are dropped by our waysides, and in midsummer the spike-like heads of this grass appear, rigidly erect, and armed with straight awns which are sometimes half a foot in length   This grain, celebrated by Pliny who called it the most ancient food of older days, is still the most important cereal of the far North, and may be raised nearer the Arctic circle than any other grain, with the exception of rye

The early Britons cultivated barley and held barley bread in high esteem, but since a statute in the reign of Edward II ordered that, "considering that wheate made into malte is much consumed, ordayned that henceforth it should be made of other graine," barley, under force of this ancient edict, has come to be the great brewing grain, and little is now heard of "bannocks o' barley meal"

The Squirrel-tail Grass, a most unworthy relative of so useful a grain, has reversed the usual order of migrating plants and for a number of years has been trav- elling eastward from the Middle States and the West. It spreads its bristly flowering-heads in waste grounds and invades dooryards and gardens, a weed wherever it appears, and furnished with that facility in transporting itself which the majority of weeds possess. Those virtues which the optimistic philosopher is so sure exist in every plant, are as yet un- discovered in this grass, and beautiful as the plant is in bloom, with its squirrel-tails of bearded spikes, its cultivation for ornament is soon abandoned. It is a slender grass, bloom- ing in early summer, and recognized by the many long awns which spread stiffly from the spike. These shining awns, often tinged with rose and lavender, are of great beauty and glisten with metallic lustre. According to a report of the Bulletin of the Torrey Botanical Club these awns perform a quite different office from such awns as those of Sweet Vernal-grass and other grasses. The awns of Squirrel-tail Grass show a backward curving which, wedge- like, raises each spikelet from those below and soon separates the ripened spike, joint from joint. The awns also, like those of certain other grasses, cling to passers-by and thus secure free transportation for the seeds.

The Wall Barley (*Hórdeum murìnum*), whose partiality for growing by walls gave to the plant its common name, is a native of Europe, and in this country is infrequently found in waste

Squirrel-tail Grass
Hordeum jubatum

places. It is a tufted annual, bearing looser sheaths, narrower, more compressed spikes, and larger spikelets than does the Squirrel-tail Grass, but its presence renders hay fully as valueless since the sharp awns, like those of the more common species, penetrate the flesh of sheep and cattle, and occasionally cause death. An English botanist recorded his earlier achievements in science when he wrote of this grass: " In our youth we put inverted spikes of the Wall Barley up our sleeves and found them travel to our shoulders. This was caused by the parts of the spikelets being compressible, so that by a gentle motion they progressed upward with a kind of spring; but the barbs, on pulling the spike the contrary way, stuck into the clothes and could not easily be dislodged."

Squirrel-tail Grass. *Hórdeum jubátum* L.

Perennial.

Stem 9'-30' tall, slender, erect. Sheaths smooth. Ligule very short. Leaves 1'-6' long, 1"-2" wide, flat, rough on margins.

Spike 2'-5' long, cylindrical, densely flowered. Spikelets 1-flowered, usually in 3's, flower of middle spikelet perfect, lateral spikelets imperfect. Rachilla prolonged. Outer scales awn-like, spreading, 1'-2½' long; flowering scale of perfect flower terminating in a slender, rough, spreading awn 1'-2' long, lateral spikelets short-awned; palets nearly as long as flowering scales. Stamens 3.

Cultivated lands and waste places, also in saline soils. June to August.

Labrador to Alaska, south to New Jersey, Colorado, and California.

Cultivated Barley

240

TERRELL-GRASS OR WILD RYE (*Elymus virginicus*). Plant was three feet high. Spikelets enlarged by two-

NODDING WILD RYE (*Elymus canadensis*).   One half natural size.   Spikelet natural size

SEA LYME-GRASS, TERRELL-GRASS, SLENDER WILD RYE, AND NODDING WILD RYE

Members of this genus have proved their usefulness in many ways The stems have been used for thatching and have been formed into a coarse fabric, the seeds have furnished an article of food to primitive tribes, and even so long ago as the eighteenth century a saline species with extensively creeping rootstocks was cultivated in Europe to preserve the shifting sands of northern coasts

In the reign of William III, the Scottish Parliament passed an act for.the preservation of Sea Lyme-grass (*Élymus arenàrius);* later, in the time of George I, the British Parliament extended the operation of this law to the coasts of England, and made it a penal offence for a person to cut the grass or to be found in possession of it within eight miles of the coast This species, the Sea Lyme-grass, is found in America, but only on the colder shores, where it is as valuable as the Marram Grass which it somewhat resembles

The eastern species of the genus are of comparatively little value and we find them chiefly in the moist soil of river banks and by low thickets, where in early summer the stout green spikes rise, stiffly bearded with upright awns, and in appearance suggesting the flowering-heads of certain cultivated grains. Terrell-grass, the one most frequently found, is the least attractive of our three common species of the genus This grass is from two to four feet tall and may be recognized by its coarse, erect spikes which are rigid and bear shorter awns than do the other species Slender Wild Rye (*Élymus striàtus*) is much more delicate in appearance, and the spikes, usually less than four inches in length, resemble a small growth of the beautiful Nodding Wild-Rye (*Élymus canadénsis*) which during the summer months ornaments wayside thickets The stout stems of Nodding Wild Rye are from two to five feet in height and bear dark green, elongated spikes which become nodding as the blossoms open

The outer, empty scales of certain species of this genus are thick and corky, and by adhering to the ripened spikelets act as floats to buoy the seeds as they fall on the water's surface This formation of the scales, so advantageous to the new seed, is most noticeable in Terrell-grass, whose spikelets, supported by their spongy floats, drift downstream until the little rafts are washed ashore and the seeds find soil on which they may take root far, often, from the parent plant.

Terrell-grass. Virginia Wild Rye. *Elymus virginicus* L.

Perennial.

Stem 2-4 ft tall, erect, rather stout. Ligule very short. Leaves 4'-14' long, 3"-8" wide, flat, rough, deep green, sometimes downy on upper surface.

Spike 2'-7' long, base usually included in loose upper sheath Spikelets 2-3-flowered, in pairs on alternate notches of the rachis Outer scales narrow, thick, and rigid, terminating in rough awns, outer scales 8"-13" long including awns, flowering scales about 4" long, usually smooth, terminating in a rough awn 3"-10" long, palets nearly as long as flowering scales Stamens 3, anthers pale yellow

Moist soil, by streams and borders of thickets. June to September.

New Brunswick to Ontario and Minnesota, south to Florida and Arkansas.

BOTTLE-BRUSH GRASS

While the Red-top is pressing the warmth of its colouring into every conspicuous place, the cool woodlands hold a few strikingly individual grasses that are not found mingling with the bourgeoisie of the fields. Shade-loving grasses of the woods are rarely crowded, and appear to be careless of that striving for position which keeps the grasses of the open pressed so closely leaf against leaf.

With the name of Bottle-brush Grass in mind this plant is instantly recognized when seen, since the loose, spreading spike is so unlike the flowering-heads of other grasses, even those of other long-awned species About this grass there is ever a suggestion of the aristocrat, none of the beggars for a roothold is this, but a plant that condescends in using the earth, and confers a royal favour by appearing in the shadows where the sunlight falls in broken gleams.

The tall stems of Bottle-brush Grass rise from among the rocks where there seems no earth in the crevices to support life, and as the pale spikelets open and spread their silvery awns the plant is one of rare beauty, worth many a long tramp in the search for it. The nodes of the leafy stems are very dark, and the lower sheaths are frequently tinged with purple The thickened bases of the spikelets are banded with narrow lines of brown, marking the place of abortive scales which in the lower spikelets appear as thread-

248

Terrell-grass. *Elymus virginicus*

Bottle-brush Grass. *Hystrix patula*

249

like awns, and in the upper spikelets show as tiny points   The ripened spikelets soon fall from the ribbon-like rachis, but the faded stems endure winter's cold and remain standing through a second season

Bottle-brush Grass. *Hÿstrix pátula* Moench.

Perennial

Stem 2-4 ft tall, rather slender, erect   Ligule very short   Leaves 5'-10' long, 3"-6" wide, flat, roughish, downy on upper surface

Spike 3'-7' long, not densely flowered   Spikelets 2-4-flowered, spreading, in 2's or 3's at each joint of the flattened rachis   Outer empty scales awn-like, sometimes 9" long, usually present only in lower spikelets, flowering scales 4"-6" long terminating in a slender, rough awn 10"-18" long, palets nearly as long as flowering scales.   Stamens 3, anthers yellowish green

Rocky woods.   June to August

New Brunswick to Ontario and Minnesota, south to Georgia and Arkansas.

BOTTLE-BRUSH GRASS (*Hystrix patula*).   One half natural size.   Spikelets enlarged by two and a half

THE SEDGE FAMILY

THE SEDGE FAMILY

CYPERÁCEAE

"This common field, this little brook —
What is there hidden in these two?"

MEMBERS of this family are protean in form, some, rising
leafless, are like green bayonets and are tipped with cylindrical
heads of blossoms, others, broad-leaved and spreading, seem like
exotics from tropical lands; some are tiny plants, rising but a few
inches from the soil, while others, stout and erect, are higher than
one's shoulder and bear great flowering-heads of innumerable
spikelets

The study of these plants is most interesting, and indeed it is
impossible to study the common grasses without gathering a large
number of these "grass-like" plants which, however, refuse to be
included with the grasses, and often prove confusing to the student
unless he has a definite idea of the distinguishing characteristics of
each family.

Resembling the green grasses in colour, sedges are most frequently spoken of as "grass," though they belong to a separate
family and show their distinctive traits in flower and growth, the
three-ranked leaves easily referring each sedge to its proper family,
since in all grasses the leaves are but two-ranked upon the stems.

The basal portion of each sedge leaf encloses the stem, as in the
grasses, but with this noticeable difference, that the sheaths of
sedges are perfectly closed while grass sheaths are usually split on
the side of the stem opposite the leaf. With few exceptions the
stems of sedges are solid, and in many species are sharply triangular The flowers are small and are arranged in spikelets, but
instead of several scales enclosing each flower, as in the grasses,
each sedge blossom is protected by but a single scale, though a
perianth is sometimes present in the form of small bristles

Many genera are comprised in the family, but the most numerous species are found among the sedges that belong to the genus
Carex, and these in northern countries equal the grasses in number

CYPERUS. *(Cypèrus)*

The sedges of this genus have been known under different names, and as "Galingale," "Earth-nuts," and "Bulrushes," many species have served the world since the days of remote antiquity. The far-famed Papyrus of the Nile is a Cyperus whose many uses it is unnecessary to recount, and in Isaiah we note the

CYPERUS HYSTRICINUS. Natural size. Spikelets natural size

CYPERUS DIANDRUS. First three illustrations natural size. [illegible] enlarged for those in illustration at lower right.

mention of sedges as we read that "the land shadowing with wings" sent ambassadors, "even in vessels of bulrushes upon the waters."

The nut-like tubers of certain sedges of this genus are edible, and the roots of a few species are fragrant and aromatic, yielding agreeable perfumes.

Our common species grow in clumps, most frequently in moist places, and bloom during midsummer and later. The stems are leafy at the base and are triangular. The flowers are borne in spikelets which are usually flat and linear and are clustered on branches at the summit of the stems. A conspicuous terminal flowering-head is thus formed, of which a distinguishing feature is the presence of one, several, or many leaves surrounding the base of the flower-cluster.

A typical plant of this genus is the Bristle-spiked Cyperus (*Cypèrus strigòsus*), a species that is common in moist or dry soil by the waysides and is also frequent near cultivated grounds, where it is often found with Edible Cyperus, or Yellow Nut-grass (*Cypèrus esculéntus*), which it somewhat resembles, and which is noted for the small, edible tubers borne on the roots. This latter species, also called "Chufa," is cultivated in southern Europe for the nut-like tubers which are said to have a sweet taste when boiled or roasted. The spikelets of the Bristle-spiked Cyperus are of greenish straw-colour, but in several other members of the genus the spikelets are noticeably coloured in stripes of brown and yellowish green.

Bristle-spiked Cyperus
Cyperus strigosus

269

On dry hillsides and in sterile soil we often notice a small, slender plant bearing globose flowering-heads of dull greenish brown. This is the Slender Cyperus (*Cypèrus filicúlmis*) whose wiry stems rise from hard, roundish corms, or tubers. The Slender Cyperus is common throughout the country, and although under different conditions of soil and climate the plants vary in size and in the number of flowering-heads the stems are seldom more than fifteen inches in height.

POND SEDGE. (*Dulíchium*)

The Pond Sedge (*Dulichium arundinàceum*) is a plant that bears little resemblance to other members of the large family of *Cyperaceae*, and the casual observer who assumed the leafy stems to belong to some flowering plant of a different order might easily be pardoned. The hollow, jointed stems, one to three feet tall, are very leafy, but the three-ranked leaves are short, being one to four inches long, and are not sedge-like in appearance. The flowers are borne with the leaves along the stem and are in spikes which are composed of narrow, green spikelets, one half to one inch long.

This plant, the only species of the genus, is common from Nova Scotia to Florida, and during midsummer and later it is frequently seen by the borders of ponds and streams where it grows with the yellow loosestrife and other plants of the marshes.

Slender Cyperus
Cyperus filiculmis

270

Pond Sedge
Dulichium arundinaceum

Slender Spike-rush
Eleocharis tenuis

THE SPIKE-RUSHES. (*Eleócharis*)

Shining, hair-like stems of Spike-rushes often cover the soil between clumps of coarser sedges, and in many places the Large Spike-rush occupies low ground by streams and ditches where its roots are sometimes under water. The smallest species are but a few inches in height, but the largest Spike-rushes are occasionally five feet tall. All are similar in appearance and are very unlike other sedges The round or four-angled stems are slender, usually rather soft or weak, and grow closely together, and the leaves are reduced to basal sheaths tinged with reddish brown. The flowers are always borne in a small, solitary spikelet which caps the stem, and in some species the spikelet is so narrow as to be no wider than the stem itself. Under the lens a perianth of bristles is noticed, and the triangular or roundish seed is seen to be tipped with the persistent base of the style.

Slender Spike-rush (*Eleócharis ténuis*), is very common in open marshes from Canada to the Gulf, and even before the plant blooms the soft, dark green, hair-like stems may be recognized as they glisten in the sunlight and sway with the slightest breeze With the Slender Spike-rush, which is usually about a foot in height, the smaller species are often seen forming loose mats above the mud, and in shallow water or in the edges of ponds the round, erect stems, two to four feet high, of the Large Spike-rush (*Eleócharis palústris*) are common

SAND-MAT. (*Stenophýllus*)

In walking along railway tracks one may often find on the embankments a rich harvest of flowers that are less common elsewhere. Even on the road-bed beneath the cars or in the sand between the tracks many sturdy little plants find place to grow and to mature seeds amid seemingly the most adverse conditions. The Sand-mat (*Stenophýllus capillàris*) is one of the low-growing plants found in such locations, and between railway ties the tufts of dark green thread-like stems capped with blackish green spikelets are frequently common The plant also grows in moister places but wherever it is found it is always low and slender, rarely a foot in height, and usually rising but a few inches from the soil

FIMBRISTYLIS. (*Fimbristylis*)

The Slender Fimbristylis (*Fimbristylis Fránkii*), and a closely related species (*Fimbristylis autumnàlis*), are small, grassy plants

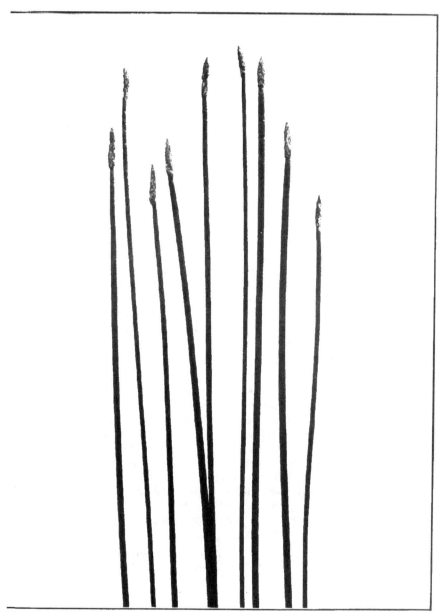

LARGE SPIKE-RUSH (*Eleocharis palustris*) Natural size

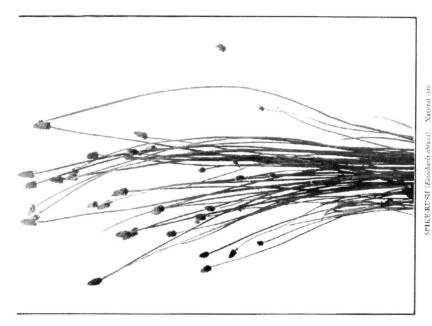

SPIKE-RUSH (*Eleocharis obtusa*). Natural size

SLENDER SPIKE-RUSH (*Eleocharis tenuis*). Natural size

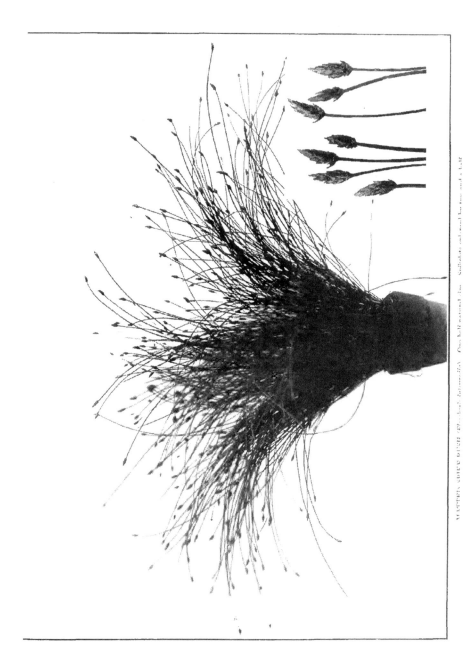

MASTER SHED RUSH (Cladium Jamaicense). On half natural size. Spikelets enlarged two and one-half

Sand-mat
Stenophyllus capillaris

Slender Fimbristylis
Fimbristylis Frankii

279

that bloom in midsummer and are usually found in moist soils. The stems are low and slender (three to sixteen inches in height) with narrow leaves and very narrow greenish brown spikelets borne on the slender branches of the umbels There are no bristles surrounding the triangular whitish seed, and as the spikelets are not clustered, but are borne in loose terminal umbels, these sedges can hardly be mistaken for others that grow in similar locations

CLUB-RUSHES and BULRUSHES. (*Scirpus*)

"Can the rush grow up without mire?"

The sedges of this genus are usually known as "rushes," and are common in shallow streams, in swamps, and in marshes   The species vary greatly in appearance, some are low and slender, being but a few inches in height, others are tall, leafless, and rush-like, while still others are broad-leaved and bear conspicuous flowering umbels.

Several of these plants were long ago noticed in homely arts Mats and ropes have been made of Bulrushes, and in early colonial days chair-bottoms of beautiful workmanship were fashioned of the Chair-maker's Rush (*Scirpus americànus*) and the Great Bulrush (*Scirpis válidus*), plants which are common in shallow water and by the borders of ponds throughout North America.

Chair-maker's Rush, recognized by its stiff, triangular stems, is found in salt-water marshes and also by inland streams   The stems, often shoulder-high, bear one to three leaves, and the flowers are borne in one to seven oblong, brown spikelets about one half an inch long   Although the cluster of spikelets is terminal it appears as if it were lateral, since the solitary leaf at the base of the cluster rises like a continuation of the stem.

The Salt-marsh Bulrush (*Scirpus robústus*) is a striking plant of the genus and is found most frequently near the coasts.   The stout, sharply angled stems are from one to five feet tall and bear a dense, compact inflorescence composed of a cluster of five to twenty large, oblong, brownish spikelets, some of which are sessile while others are borne on short rays one to two inches in length

The Great Bulrush, occasionally nine feet in height, and sometimes an inch in diameter at the base, is leafless and stout.   The

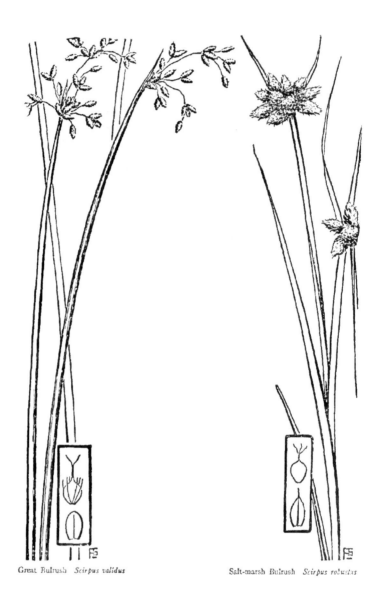

Great Bulrush *Scirpus validus*   Salt-marsh Bulrush *Scirpus robustus*

Wool-grass *Scirpus cyperinus*   Meadow Bulrush *Scirpus atrovirens*

WOOL-GRASS (*Scirpus cyperinus*) growing in marsh. River Bulrush (*Scirpus fluviatilis*) growing around the Wool-grass

IA LAKE SHORE. Chair-maker's Rush (*Scirpus americanus*) and Great Bulrush (*Scirpus validus*) growing in foreground

CAYUGA MARSH. The cove where grow Wild Rice, Blue-joint Grass, Wool Grass, River Bulrush, and many sedges

RIVER BULRUSH (*Scirpus fluviatilis*)   tural size
Fruit spikelets natural size

MEADOW BULRUSH (*Scirpus atrovirens*). One half natural size

SALT-MARSH GRASS, II.

SALT MARSH BULRUSH (*Scirpus robustus*). About three quarters size

A THICKET OF SALT-MARSH BULRUSH (*Scirpus robustus*) by the edge of the marsh

4

WHITE BEAKED-RUSH ......
Spikelets enlarged by two

.... (.............. virginicum).
Natural size

smooth, round stems are often seen in shallow brooks and are more noticeable than are the small terminal umbels of brownish spikelets.

Wool-grass (*Scírpus cyperì-nus*), common in late summer, is a handsome plant of low meadows, where groups of this rigid, leafy sedge show a rank growth. The stems of Wool-grass are frequently shoulder-high and bear a profusion of rather narrow, long, drooping leaves which are sharp-edged with minute saw-teeth. The perianth, which in this genus is present in the form of bris-tles, is in Wool-grass long and downy, clothing the spikelets of the conspicuous terminal umbel in dull gray wool.

The Club-rushes (*Scírpus polyphýllus* and *S. sylváticus*) are leafy and graceful plants that are frequently noticed in low meadows and in swamps. They are usually from two to six feet tall and bear broad leaves and large terminal um-bels of many small spikelets. These flowering-heads are coloured in dull tones of green and brown, and we must push aside the overshadowing leaves and the low masses of surrounding vegetation in order to find the striking tints of deep carmine, corn-colour, and pale rose which are frequently concealed at the base of stem and leaf.

Slender Cotton-grass
Eriophorum gracile

Virginian Cotton-grass
Eriophorum virginicum

COTTON-GRASSES. (*Erióphorum*)

Where the arethusas and pogonias hide their fragrant flowers among low sedges the Slender Cotton-grass (*Erióphorum grácile*) rises above the smaller growth, and to the purple of the wild iris and the deep colour of the large blossoms of pitcher-plants this sedge offers the contrast of flowers clothed in shining white

The Cotton-grasses — in reality Cotton-sedges — are most easy of recognition, since in the common species the blossoms at maturity are surrounded by long, silky hairs

Slender Cotton-grass blooms at least a month earlier than the Virginian Cotton-grass (*Erióphorum virgínicum*), and as the level light of sunset lies across the meadows of late spring the plumes of this earlier species glisten with added lustre, and the marsh which was so desolate in winter, so golden with marsh marigolds in April, is abloom in frosted silver

The Virginian Cotton-grass, from one to four feet in height, is common from Canada to Florida In this larger species the spikelets are more densely clustered than are the spikelets of Slender Cotton-grass, and the hairs composing the perianth are usually rusty brown or tawny in colour

Certain species of this genus have been used in interesting experiments in which thread has been spun and firm cloth made from the silky hairs of the spikelets, but the fibre is short, and the attempt to bring the hairs into use as a substitute for cotton was soon abandoned Where Cotton-grasses grow in abundance on the moorlands of Scotland the poorer people formerly found a use for the down in making candle and lamp wicks

BEAKED-RUSHES. (*Rhynchóspora*)

Sated with the brilliancy of Fire-weed and flaming lilies one turns to search for less commonly known plants, and then one may notice in low grounds and near the borders of swampy thickets, slender stems of the Beaked-rushes bearing flower clusters of white, green, or rich brown crowned with white stigmas and pale anthers A perianth is usually present in the form of rough bristles, and aside from the more noticeable characteristics of growth the plants of this genus may be recognized by the beaked seeds to which the common name refers

Beaked-rushes are common during midsummer, and even

though they may be grass "to the general" there is something in
the appearance of the small flower clusters that excites curiosity
and urges that botany and micro-
scope be brought into use. In
Clustered Beaked-rush (*Rhynchóspora glomeràta*) the leaves are dark
green, narrow, and erect, and several
clusters of dark brown pointed spikelets are borne at intervals along
the stem. White Beaked-rush
(*Rhynchóspora álba*), a smaller and
more slender species than the preceding, is also common in moist
grounds. The leaves of White
Beaked-rush are light in colour, and
the few flower clusters borne near
the summit of the stem are of
pure white.

NUT-RUSHES. (*Sclèria*)

The Nut-rushes are small, slender sedges, not uncommon in
marshes and low meadows during
midsummer, although seldom noticed among the taller growth that
surrounds them. Low Nut-rush
(*Sclèria verticillàta*) is very slender,
never more than three feet in
height and usually much less than
that. The stems are sharply three-
angled and bear a few narrow leaves
above which are the small spikelets
in four to six sessile, green or
purplish clusters. The ripened fruit
is more conspicuous than are the
flowers, as the shining white seed
is very prominent. The generic
name, derived from the Greek, al-
ludes to the hardness of these
"nuts," which are roughened by

Clustered Beaked-
rush. *Rhynchospora
glomerata*

Low Nut-rush
Scleria verticillata

303

Perigynia

seed with 3 stigmas

seed with 2 stigmas

scales

broken horizontal ridges and well repay a close examination with the microscope.

SEDGES. (Càrex)

Sedges of this widely distributed genus grow in abundance in wet meadows, by brooksides, and in all swampy places. In a rich locality more than fifty species may be gathered in the course of a summer, and to the student these plants are more perplexing than are the grasses, since between many of the sedges the difference is too slight to be obvious save to patient study aided by good lenses.

"Shear-grass" is an old English name for the sedges, and one that is most appropriate, since the leaves and stems of many species are sharp-edged and can cut sorely if carelessly handled.

The plants of this genus are perennials, growing in tufts, and sending up many long-leaved, sterile shoots which, like the flower-bearing stems, are three-angled at the base. The flowers are borne in spikes in which the staminate and pistillate flowers are variously arranged: some sedges show spikes composed partly of fertile and partly of staminate blossoms, while in other sedges whole spikes are of one form of flower. The difference between the two forms is especially noticeable as the seeds ripen. The staminate spike is narrow and slender, but the pistillate spike increases in thickness with the ripening seed. When

Showing different forms, scales, seeds, and perigynia in sedges belonging to the genus *Carex*

Tufted Sedge
Carex stricta

304

both forms are borne in the same spike that part which bears staminate flowers is slender and contracted, and the pistillate part, which is sometimes at the base of the spike, sometimes in the middle, and occasionally at the summit, is swollen with seed, thus giving a ragged, uneven appearance to the inflorescence. The seed is enclosed in a sac, known as the perigynium, and the form of the seed is determined by the number of stigmas; when there are two stigmas the seed is more or less compressed, or two-edged, while with three stigmas the form of the seed is triangular. Each flower is protected by a scale, or bract, and a close observance of the form of the scale is of the utmost importance in determining the species.

Brilliant colours are seldom found in this genus; the rose and purple which adorn the flower of the grass are lacking, and the blossoming spikes of sedges are of dull green or brown, while rarely is the base of the stem tinged in reddish.

Our first acquaintance with sedges is usually gained by using the stout clumps of Tufted Sedge (Càrex stricta) as stepping-stones while crossing a wet meadow. This sedge is one of our most common species, and the tufts of long, gracefully spreading leaves should be well known to those who search the swales for orchids and pitcher-plants, since the firm tussocks save one many an undesired plunge into muddy water. The Tufted Sedge blooms from June till August and is very common by damp waysides and in wet soil, where the plant is usually about three feet in height with rough, three-angled stems at whose summits are borne three to five narrow, erect spikes of blossoms. The spikes are

Pennsylvania Sedge
Carex pennsylvanica

307

from one to two inches long, with deep brown scales which show green mid-veins and lighter coloured margins.

The earliest of our common species, as well as one of the most abundant, is the Pennsylvania Sedge (*Càrex pennsylvánica*) which is found in the dry soil of hills and open woodlands, carpeting the ground with tufts of slender leaves and opening inconspicuous flowers before the more showy blossoms have wakened from their winter's sleep. In the borderland between woods and open pasture this sedge is one of the most common plants, and by dry waysides it is also frequently seen. The stems are rarely more than a foot in height and are often much less than that. The blossoming spikes are small, the staminate flowers being borne in the uppermost spike, which is from one half to one inch in length, while the fertile flowers are in smaller, sessile spikes immediately below the staminate blossoms. The scales are dark reddish brown, lighter on the margins and along the mid-vein.

Bladder Sedge

Carex intumescens.

Hop Sedge *Carex lupulina*

Fringed Sedge
Carex crinita

The most noticeable sedges of open marshes are the several Hop Sedges, blooming in early summer and bearing thick, oblong spikes of inflated, light green seed-pouches. Of the common species the Bladder Sedge (*Càrex intuméscens*) is a slender plant with one to three short, few-flowered, fertile spikes above which the narrow staminate spike is borne on a slender stalk. The Hop Sedge (*Càrex lupulìna*) is stout, with broad, light green leaves and two to six densely flowered fertile spikes which are usually sessile, though the lower spike is often borne on a short peduncle. The Porcupine Sedge (*Càrex hystricìna*) bears narrow, yellowish green leaves and one to four densely flowered pistillate spikes. The long, rough point of the scale is a distinguishing feature of this species. The Long Sedge (*Càrex folliculàta*) is tall and slender and is most frequently found in the borders of thickets and swamps. It should be recognized by the broad, flat leaves and the rather loosely flowered pistillate spikes which are borne on long, spreading peduncles.

The most showy sedge of swamps and open woods is the great Fringed Sedge (*Càrex crinìta*) which grows shoulder-high, with sharp, three-angled stems which spread from the

Little Prickly Sedge
Carex scirpoides

Slender Sedge
Carex gracillima

310

REX CRISTATA. Natural size   CAREX LURIDA. One half natural size   CAREX MIRABILIS Natural size

HOP SEDGE (*Carex Pseudo-Cyperus*). One half natural size. Fruit spike natural size.

CAREX LUPULIFORMIS. One half natural size. Spike natural size.

FRINGED SEDGE (*Carex crinita*). One half natural size

clumps and rest against the alders and the low shrubby growth of the thickets. Indeed, if there be truth in the old couplet concerning

"The alder whose fat shadow
nourisheth,
Each plant set near to him
long flourisheth;"

the Fringed Sedge, with its rank growth, would seem to have derived benefit from that "fat shadow." The plant blooms in midsummer, and at that season the long, drooping spikes are very noticeable with their spreading, long-awned, brownish green scales. The staminate spikes are sometimes fertile at the base or in the middle, while the pistillate spikes are frequently staminate at the tips.

Another common and easily recognized sedge of low grounds is the Little Prickly Sedge (*Càrex scirpoìdes*) which blooms during the spring and summer. The leaves are thread-like, and the slender stems, seldom rising more than fifteen inches from the soil, form low tufts of light, glistening green. The three to six tiny spikes are very short and are almost globose in form and as they open widely appear like small, star-shaped flowers.

Slender Sedge (*Càrex gracíllima*) is common in moist meadows and also in drier soil by the waysides, where this plant is frequently found in the low, hedge-

Fox Sedge
Carex vulpinoidea

Fescue Sedge
Carex festucacea

321

like growth that borders country walls and fences. The whole plant, stem, leaf, and flowering-spike, is slender and bends to the slightest breeze This sedge may be recognized by the green, fertile spikes which are very narrow and droop from hair-like peduncles

Fox Sedge (*Càrex vulpinoìdea*) is common in low grounds and by waysides, and is not infrequently found in dry soil The stout stem is sharply three-angled and bears short, green or brownish green spikes, which are densely crowded in clusters and form what appears to be a rough, terminal spike two to five inches long

The Fescue Sedge (*Càrex festucàcea*) belongs to a group in which there are several species that bear more or less resemblance to one another In these sedges the short, densely flowered spikes are round or oblong and are borne rather closely together at the summit of the stem. Fescue Sedge blooms in spring and early summer and is most frequently seen in dry soil.

The above species of the genus *Carex* are but a few of the most common sedges that may be found in many locations from Canada to the Gulf. The majority of these plants bloom during spring and summer, although many of them retain their ripened seeds during the autumn months. The sedges, no less than the grasses, have developed many interesting peculiarities of structure which aid in transporting the seeds to new fields a few of the sedges growing in dry soil show wing-margined seed-pouches that are easily carried by the wind, some bear rough pouches that catch on passers-by and are carried far, sedges of marshlands often float their seeds in inflated sacs along the water's surface or buoy them by corky growths to which the base of the seed is attached None of Nature's children are so lowly that she neglects to prepare them for the world-long struggle for existence, and the methods which these humble forms have developed in order to enter into successful competition with the crowding life around them show the restless energy of living matter in its effort to exist and to perpetuate itself in life.

RUSHES

YARD RUSH (*Juncus tenuis*)   Two thirds natural size

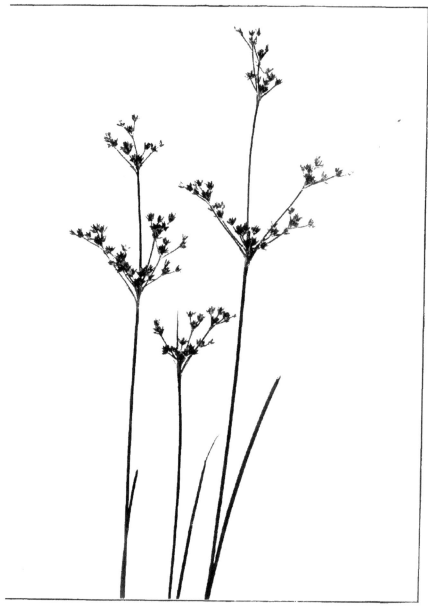

SMALL-HEADED RUSH (*Juncus brachycephalus*). Natural size.

RUSHES

JUNCACEAE

"Where is this stranger? Rushes, ladies, rushes!
Rushes as green as summer for this stranger"

THE dark-green leaves of rushes, common in lanes, by waysides, and in all marshy places, add their verdure to the green carpeting of the grasses, and indeed are often familiarly comprised under the all-embracing name of "grass" Rushes, however, belong to a distinct family, and, being closely related to the lilies, are essentially lily-like in the form of the tiny green blossoms.

In song and story we are familiar with poetical allusions to "the gleaming rushes," yet many of the so-called "rushes" belong to the sedge family, and are far larger than our common Eastern species of the true rushes and seem to a passing glance fully as rush-like in appearance While it is true that plants of this genus, (*Juncus*), resemble both grasses and sedges in colour and texture, the student should have no difficulty in distinguishing them, since each flower of the rushes, like a miniature lily, shows a perfectly six-parted perianth, three to six stamens, and three stigmas

The largest of our common species is the Bog Rush (*Júncus effúsus*) which grows in clumps in moist places and is so often seen by waysides and in low meadows The round stems, stiffly erect, are from two to four feet in height, and are filled with soft, white pith The stem is leafless, and several inches below the pointed tip hangs a many-flowered cluster of green blossoms which seems to have burst from the stem, though, in reality, the upright portion above the blossoms is a leaf of the inflorescence, and as this leaf appears as a continuation of the stem it causes the flower cluster to seem lateral The Bog Rush is one of the few meadow plants that remain green until late autumn, and even in winter we may often notice low tufts of the dark-green stems by winding brooks

The Jointed Rushes are peculiar in that the interior hollow portion of their leaves is divided by horizontal, membranous par-

Bog Rush
Juncus effusus

Sharp-fruited Rush
Juncus acuminatus

The peculiar growth of
the spikelets shown at
right in illustration is frequently found in this rush

Grass-leaved Rush
Juncus marginatus

titions which, as joints, or knots, may be plainly felt when a leaf is drawn through the hand. The leaves in the majority of the species of Jointed Rushes are round. In the Sharp-fruited Rush (*Júncus acuminátus*), a species which is usually from one to two feet in height and is very common in bogs, the inflorescence is composed of spreading terminal branches tipped with small, closely flowered heads. The narrow divisions of the perianth are sharp-pointed and are reddish brown in colour.

The Grass-leaved Rush (*Júncus margináta us*) is found in moist sandy places. The stem, seldom more than two feet tall, is erect and somewhat flattened, and, as the common name indicates, the leaves are long, flat, and grass-like. The inflorescence is composed of three to twenty small, brownish green heads of flowers. There are but three stamens and the anthers are reddish brown in colour.

In the following species the flowers are placed singly on the branches of the inflorescence and are never in true heads. The leaves are grass-like.

Yard Rush (*Júncus ténuis*), common in country dooryards and by footpaths, seems to thrive best when it is trodden under foot each day. This rush grows in low-spreading clumps of wiry, glistening stems which are leafless except at the base from whence numerous narrow leaves rise. The leaves are shorter than the stems, but the inflorescence is much exceeded by the lowest involucral leaf which is usually from three to seven inches long. Through June, July, and August the plant is in bloom and the tiny flowers, scattered along the branches of the inflorescence, or crowded at their tips, are like pale stars. The perianth, green on its outer surface, is whitish within, and the six short anthers and the feathery

Yard Rush
Juncus tenuis

stigmas are also white. The flowers are widely open in the early hours of the morning but close during the heat of the day.

The Toad Rush (*Júncus bufônius*) is an odd little plant, rarely eight inches tall, which often spreads in tangled mats over low ground by the waysides and on the borders of dried-up pools. The flowers are larger than in our other common rushes, and are dark green; the stems branch abundantly at the base and bear one or two short, narrow leaves. Like the preceding species the Toad Rush is found throughout nearly the whole of North America.

Black-grass (*Júncus Gerárdi*), easily recognized by its characteristic dark-green colour, blooms in midsummer and is common along the Atlantic coast and by tidal waters of rivers from Canada to Florida. The plant is grass-like, and with dark leaves and blackish flowers covers large areas on the salt marshes, where it is often associated with Foxgrass (*Spartína pàtens*). The slender wiry stems of Black-grass rise from creeping rootstocks and are usually from one to two feet in height; the perianth divisions are rounded and are shorter than the dark seed-capsule. This rush is the most highly valued of the common species, as it yields a large part of the salt hay that is taken each year from our coastwise marshes.

Other rushes will occasionally be found by the student and may

Toad Rush
Juncus bufonius

SHARP-FRUITED RUSH (*Juncus acutiflorus*). Natural size.

BULRUSH (*Juncus effusus*). One third natural size.

OTTED RUSH (*Juncus nodosus*). Two-thirds natural size   JOINTED RUSH *Juncus articulatus* half natural size

be distinguished by the general manner of growth, the form of the small divisions of the perianth, and the relative length of these divisions in comparison with the length of the seed-capsules.

The gathering of rushes was an important task when the floors of English dwelling houses were covered with these plants of the marsh, and the sovereign could require, as did William the Conqueror of his subjects upon Aylesbury land, that the people furnish "straw for his bedchamber . . . and in summer straw rushes." To this floor-covering Erasmus ascribed pestilences, since the lowest layer of rushes was often left unchanged for years. In the days of "Merrie England" such rush-strewn floors were an evidence of barbarism to the courts of southern Europe, where a Frenchman of the eighteenth century reported to Henry III of France that there were but three remarkable things to be seen in England, of which one was the custom of the people to "strew all their best rooms with hay." We also read that in olden days the pathways of processions were made green

Black Grass
Juncus Gerardi

Common Wood-rush
Luzula campestris

337

with scattered rushes, and that in Shakespeare's time the stage was strewn with these plants.

Rush-lights of bygone days were prepared from the pith of certain plants of this genus   The round stems were gathered in late summer and were placed in water for a short time   The pith was then carefully taken from the stems, and after being left out in the dew for several nights was dried, and dipped in scalding fat.

THE WOOD-RUSHES.   (*Lùzula*)

While searching for the earliest hepatica or arbutus the soft, reddish green leaves of the Common Wood-rush (*Lùzula campéstris*) are often seen   This rush is one of our earliest flowering plants, and appears while the turf still remains brown from winter's frosts   Common Wood-rush grows in tiny tufts and is found in many locations from dry, open woodlands to low marshes, and through all the summer months the plant remains noticeable as its ripening seeds bend the slender stems earthward with increasing weight   There is seldom a rocky pasture that does not show a few of the reddish umbels spreading from the low growth that so universally surrounds each firmly embedded stone, while a favourite location is near the borders of open woods, where later the Pennsylvania Sedge carpets the ground beneath white birches and low-growing oaks   When the flat, rather broad leaves first appear they are sparingly fringed with silky white hairs.   The plant is rarely more than a foot in height and the blossoming umbel is composed of short branches which bear small, densely flowered spikes

The Hairy Wood-rush (*Lùzula saltuénsis*) prefers dry, wooded banks and is distinguished from the more common species by the one-flowered, hair like divisions of the umbel, by the more numerous long hairs on the leaves, and by the perianth which differs from that of the other in being shorter than its capsule.

The generic name of the Wood-rushes, *Luzula*, is said to have been derived from the Italian word for glow-worm, and probably referred to the shining seed-capsules.

INDEX TO ENGLISH NAMES

INDEX· TO ENGLISH NAMES

The Book of Grasses

342

INDEX TO LATIN NAMES

INDEX TO LATIN NAMES

THE END

THE COUNTRY LIFE PRESS
GARDEN CITY, N. Y.

CPSIA information can be obtained
at www.ICGtesting.com
Printed in the USA
LVHW080157041121
702431LV00009B/119